THE BEACH BOOK

THE
BEACH BOOK

Science of the Shore

CARL H. HOBBS

 Columbia University Press *New York*

Columbia University Press

Publishers Since 1893

New York   Chichester, West Sussex

cup.columbia.edu

Library of Congress Cataloging-in-Publication Data

Hobbs, C. H. (Carl Heywood), 1946–

  The beach book : science of the shore / Carl H. Hobbs.

    p. cm.

  Includes bibliographical references and index.

  ISBN 978-0-231-16054-4 (cloth : alk. paper)

  ISBN 978-0-231-16055-1 (pbk. : alk. paper)

  ISBN 978-0-231-50413-3 (ebook)

  1. Beaches. 2. Coastal ecology. I. Title.

  GB454.B3H63 2012

  551.45′7–dc23

                             2011047619

CONTENTS

ACKNOWLEDGMENTS

The author of a book is more like the captain of a team than a solo operator. I could not have written *The Beach Book* without the encouragement and assistance of many individuals. The idea for the project evolved during a series of conversations with my daughter, Catherine Hobbs. Her involvement has continued as she helped edit chapters, contributed advice about the ways of the publishing industry, and encouraged me to continue working whenever the inevitable frustrations appeared. More than forty years ago, Miles Hayes introduced me to the study of coastal geomorphology and coastal processes. So, in a different sense, the book began with him. Patrice Mason drew several of the figures and provided encouragement and moral support while I worked on the manuscript. Throughout my career at the Virginia Institute of Marine Science (VIMS), I have learned from professional associations and friendships with many talented colleagues, including John Boon, Bob Byrne, Scott Hardaway, John Milliman, Jim Perry, John Wells, and Don Wright, among others. Susan Schmidt's interest, encouragement, and labors as an editor substantially improved the manuscript. The project could not have gone forward without Patrick Fitzgerald, publisher for life sciences at Columbia University Press. Bridget Flannery-McCoy helped me navigate many of the details of publishing, and Irene Pavitt brought her superb skills as an editor to the book. I thank each of them and the many others who, in one way or another, contributed to my efforts in preparing the manuscript.

THE BEACH BOOK

INTRODUCTION

The Beach Book describes the physical processes and materials that create and change the "edge of the sea." Sunbathers who visit the shore on day trips or week-long vacations notice that the beach is different each time they return. It can be wide or narrow, pitch steeply or slope gently into the water, be rocky or sandy. Swimmers feel the currents. Many popular beaches are on barrier islands with sand dunes. Marshes that feed and shelter fish and crabs often are near beaches. Anglers and boaters navigate through tidal inlets, experience the roll of waves, and must remember the hours of high and low tides. Home buyers want to gauge erosion rates. The more we know about the physical processes that create, maintain, and erode beaches, the better we can appreciate the shore. This book presents basic geology and oceanography for interested readers.

I have worked mainly along the coasts of the Atlantic Ocean and Gulf of Mexico. However, the concepts and issues covered in this book apply to coastal systems worldwide. In all areas, tidal range, sediment characteristics and abundance, wave climate, and sea-level change work together to create and shape beaches. Even though the relative importance of each factor varies from place to place, the underlying physical rules do not.

Mathematics helps explain these physical rules, and this book presents some key equations about the forces that build and change beaches. While math might challenge some readers, every equation is clear and the accompanying text describes the relationships stated in the equations.

Some readers might work through the math in the chapters on waves and tides, while others might be content with the text.

After an overview of beaches, the book considers wind. Wind can bring cold water to a hot beach in summer, and storms are named by the direction of their winds. Most important, wind generates waves, which move sand to shape beaches. While we look forward to splashing in the waves or surfing on them, waves also are the major agent of beach erosion. Similarly, tides play a major role in shaping beaches and in the dynamics of inlets between barrier islands. Sediments are the stuff of beaches, accounting for some of the differences among beaches in different locations.

The book then expands the discussion on beaches and integrates concepts to include a broader segment of the coast that is familiar to many beachgoers: the closely related barrier islands and inlets, and the very disparate sand dunes and tidal marshes. Barrier islands and tidal marshes, in particular, are very sensitive to changes in sea level. A worldwide social and political as well as environmental challenge, sea-level rise requires that not only marine scientists but also policy makers and citizens understand the geological processes that control sea level, the impacts of long-term changes in water level on coastal areas, and the projections for sea-level rise in the future. Storms and storm surges are powerful and fast-acting meteorological events with geological consequences. One of those effects is damage to the shore, including erosion. The book ends with a discussion of the tools used to resist erosion and the scientific and social factors to consider in deciding whether to combat it.

Because each chapter is written so that it can stand on its own, there is some repetition. By explaining the physical aspects of coastal systems, this book will help natural-resource managers make important decisions. Citizens will better understand public-policy options concerning the control of shoreline erosion, the restoration of eroded beaches, and the regulation of coastal-zone development.

The Beach Book will give readers a better appreciation of beaches, the areas around them, and the processes that shape the shore.

1
BEACHES

For many people, the beach is the shore. It is where they go to play in the sea, to sunbathe, to play beach volleyball, and to dance to beach music. In some uses, the words "shore" and "beach" are synonymous. An unemployed or a retired mariner is said to be "on the beach." To the coastal scientist, however, a beach is a specific, physical environment. In *Beaches and Coasts*, Richard Davis and Duncan FitzGerald define it as a "deposit of unconsolidated sediment, ranging from boulders to sand, formed by wave and wind processes along the coast. The beach extends from the base of the dunes, cliff face, or change in physiography seaward to the low-tide line."

Waves, currents, and wind move the sediments and shape the beach. The sediments at the beach are mobile, since they are both unconsolidated and *noncohesive*; that is they do not stick together. Beach sediments range in size from boulder to sand. The *cohesive* fine-grained sediments, silt and clay, do not play a significant role because they are unlikely to be deposited in the high-energy environment that helps create the beach. The shape of the beach is the response of the sediments to the physical forces of wind and water. The narrow, sandy strand is the buffer between the energetic sea and the relatively stable land. The beach changes form to accommodate fluctuations in the forces working on it. The shapes of the beach and the nearshore adjust during storms to dissipate energy. The coast responds to changes in the driving processes at a wide range of spatial and temporal scales. During a storm, the beach can change shape in an hour or less. It can take days or weeks for a beach to grow. The entire shore area can change over decades.

There is a full vocabulary of terms to describe different parts of the beach profile and adjacent areas (figures 1.1 and 1.2). Variations in the terminology are common; for example, the bar and trough sometimes are called the *ridge* and *runnel*, and if the bar / ridge grows vertically so that it is close to the water level, it sometimes is called a *swash bar*. In addition, every feature—for example, the storm or winter berm—is not always present on a beach. Conversely, a beach often has two or more ridges, each associated with a breaker zone.

The *berm* is the relatively flat, upper surface of the beach. It is the most hospitable place for sunbathing and playing, where beachgoers spread their blankets and play Frisbee or volleyball. The elevation of the berm is slightly above the level of high tide. The *beach crest*, or *berm crest*, is the break in slope that separates the beach face from the berm. It is the seaward edge of the berm and usually is easily discernible by eye. The *beach face*, or *foreshore*, is the surface that slopes down to the water. It is too steep for volleyball and is washed by wave run-up, so it is too wet for most sunbathers. The *step*, or *beach step*, is a usually small notch at the bottom of the beach face. It is commonly composed of coarser sediments than those that occur landward or seaward. Because the step is underwater, we seldom see it, but as we wade out into the water, our feet feel both the sudden, small drop and the pebbles. Those of us with tender feet quickly learn the location of the step and try to avoid it. As we walk farther offshore, the slope flattens into the *low-tide terrace*. Often the landward face of the first ridge is quite steep.

Coastal scientists sometimes refer to differences between a winter and a summer beach, but this dual nature is not universal. Some of the earlier and better-known beach studies took place on the West Coast of North America, where the winter–summer differences are strong. Compared with the coasts of the Atlantic Ocean and Gulf of Mexico, the Pacific coast has a more distinct change from summer to winter in wave regime. Along the East and Gulf coasts, hurricanes come ashore from midsummer through late autumn, and *nor'easters* (northeasters) from early

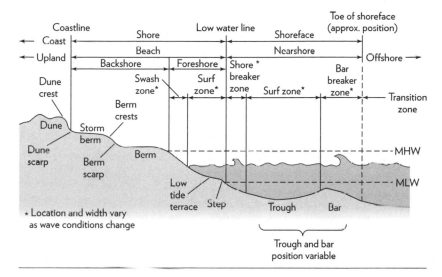

Figure 1.1 Beach terminology. (After A. Morang and L. E. Parson, Coastal terminology and geologic environments, in U.S. Army Corps of Engineers, *Coastal Engineering Manual*, EM 1110–2-1100, part 4, chap. 1 [2002], http://chl.erdc.usace.army.mil/CIIL .aspx?p=s&a=ARTICLES;104; redrawn by Network Graphics)

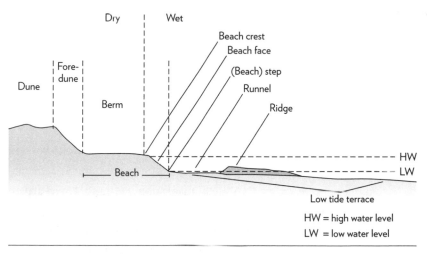

Figure 1.2 Alternative beach terminology. (Illustration redrawn by Network Graphics from an original by the author)

autumn through mid-spring, leaving generally calm conditions during late spring and early summer. There is much less seasonality to wave energy along the coasts in the eastern United States and, thus, to the shape of the beach.

There are additional terms for the along-shore shape, or planform, of the beach. While it is fairly easy to define the *shoreline*, it is more difficult to find it. The shoreline marks where the beach and the water surface meet. As the water constantly moves up and down, whether it be rapid change due to waves and swash or slower change due to tide, the shoreline changes as we watch. The high and low waterlines are also difficult to locate with precision but are more stable than the shoreline, which is good because they frequently have very real legal implications, such as marking the seaward limit of private property. The high waterline is the intersection of the shore and the horizontal plane at the elevation of *mean high water* (MHW) or sometimes *mean higher high water* (MHHW). The low waterline is the intersection of the shore and the horizontal plane at the elevation of *mean low water* (MLW) or *mean lower low water* (MLLW). These lines are stable vertically, at least through the 19-year *tidal epoch*, but move laterally as the beach advances and retreats.

Beaches in specific settings may be called pocket, or headland, beaches or barrier beaches. A pocket beach occurs between two promontories or headlands. Viewed from above, pocket beaches have a crescent shape that is the result of the waves bending (refracting) around the headlands. The beach usually is an asymmetrical arc, as waves seldom approach from directly offshore. In some places, the asymmetry shifts from season to season as the regional wave-approach direction (wave climate) changes. Barrier beaches are the seaward fringe of barrier islands or (barrier) spits. An estuary or a lagoon or sound separates the mainland from the barrier, which usually is relatively straight. One end of a barrier spit is attached to the mainland, where there is a substantial supply of sand. Wave-driven currents carry sand from the source toward the end of the spit. Hence the direction in which the spit lengthens shows the dominant, or net, direction of the longshore transport of sediment. If the distance from the end of the spit to the next segment of land, or between two barrier islands, is relatively narrow, the open water is an inlet. Wide openings often are bay or river mouths.

Immediately after a hurricane or nor'easter erodes the shore, there usually is a lot of public concern about the loss of the beach. But, frequently, a couple of weeks later, the angst fades because much of the beach has returned of its own accord. This cycle of loss and recovery is similar to the seasonal cycle that creates summer and winter beaches.

Storms erode sand from the landward portions of the beach and move it both along shore and offshore. If a storm lasts through several high tides or has at least a moderate *storm surge*, it may erode the dune front as well. If we assume that the rate of sediment transport parallel to the shoreline is the same along any particular stretch of beach, the profile becomes lower and smoother. The berm, the visible portion of the beach, shrinks because sand is eroded and moved offshore. The sand removed from the beach is deposited in the nearshore and shallow offshore regions, so the water farther offshore becomes slightly shallower than it was before the storm. It is very difficult to survey underwater while the waves are high and breaking during the storm, so measuring and quantifying the change is hard. As the beach profile changes, waves break farther away from the shoreline and dissipate their energy across the wide nearshore zone. But the main reason that storm waves break farther from the beach is that larger waves break in deeper water. At times, the surf zone is wide enough that there are multiple lines of breakers as large, storm waves first break in relatively deep water, then re-form and break again, and, perhaps, again and again. In ever-shallower water, they finally exhaust themselves on what is left of the beach face. This long, gently sloping profile is known as a *storm beach* or *dissipative beach* (figure 1.3A), since it provides ample space for the dissipation of wave energy.

As the storm passes, the wind often shifts so that it blows offshore, the storm surge ebbs, and the waves diminish. Low, calm-weather waves create bottom currents that transport sediment across the low-tide terrace and back toward the shore. As the sediment returns shoreward, it builds a small ridge on which the waves break (figure 1.3B). The ridge and breaking waves accentuate the landward transport of sediment, since the sand

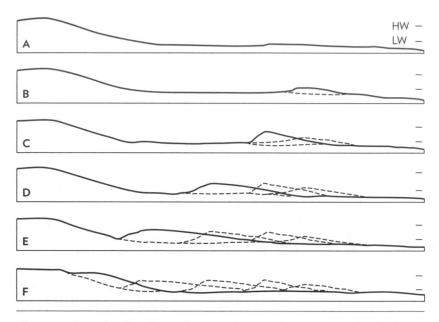

Figure 1.3 The cycle of beach growth: HW and LW are the approximate levels of mean high and mean low tide, respectively; (*A*) a dissipative beach shortly after a storm; (*B–E*) the initiation and growth of a ridge and its movement across the low tide terrace, with the parallel development of a runnel between the ridge and the beach face; (*F*) the welding of the ridge onto the foreshore and the resulting growth of the beach. (Illustration redrawn by Network Graphics from an original by the author)

cannot return seaward over the newly forming ridge. The dominantly shoreward movement molds the bar into an asymmetrical form with the steep face on the landward side. The ridge continues to migrate landward, narrowing the runnel (figure 1.3*C–E*), until it reaches and "welds" to the foreshore (figure 1.3*F*). Given the right wave conditions, the ridge can migrate up the foreshore to about the upper limit of the wave run-up. Beaches grow both vertically and horizontally through this process. More than one ridge can move across the low-tide terrace at any time.

This natural beach-nourishment process can take several days to several weeks. The maximum height of the top of the ridge above the floor of the runnel or the low-tide terrace is related to the *tidal range*. Frequently, the steepness and height of the landward face of the ridge can be great enough to impede wading. Several years ago at Ocean City, Maryland, I was in waist-deep water within arm's reach of people farther offshore, but they were in water that was only ankle deep.

Knowledge of the natural process of beach growth can have practical value. When a storm removes part of the beach face and berm, the awareness that it will rebuild over a few weeks can relieve the anxiety that a property owner or resource manager may feel. Understanding how and why an offshore bar is likely to move over the course of a few days can be a matter of life and death for a Marine Corps officer planning an amphibious assault.

A strong ridge-and-runnel system, especially when the runnel is relatively narrow, leads to a complex and dynamic nearshore region. When the crest of the ridge is close to the water level, spilling or breaking waves push water into the runnel. The water has to escape; that is, it wants to return to its own level, but the ridge prohibits a direct backflow. Thus the water flows alongshore, parallel to the shoreline, in the runnel until it reaches a low area in the ridge where it is strong enough to push seaward in a narrow jet. This strong offshore flow, often concentrated in the surface water, can cut through the surf zone until it disperses in calmer water. This narrow current flowing quickly away from the beach is a *rip current*. Lifeguards warn swimmers that rip currents can carry them seaward and teach them to swim parallel to shore rather than fight the jet. After swimming a short distance along shore, the swimmer should be out of the rip and able to return to shore without having to swim against the current.

Rip currents can occur on beaches without ridge-and-runnel systems. Waves, especially when aided by an onshore wind, push water toward the shore, where it piles up. Eventually, the difference in elevation between the water stacked against the beach face and that farther offshore, even though small, is enough to force an outflow. This tends to be a jet-like stream of water through the surf zone—that is, a rip current.

An undertow is not a rip current, although sometimes the terms are used synonymously. Occasionally, instead of forming into discrete jets, the return current is a sheet flow. The wave backwash (-swash) down the beach face continues beneath the new up-swash, causing a two-layer flow.

Figure 1.4 The complex set of bed forms in a runnel landward of a ridge at low tide. The sand waves formed as water trapped in the runnel by the falling tide flowed along the shore to escape. (Photograph by the author)

The upper layer is the remnant of the breaking wave moving landward, and the lower layer, the under tow, is the back-swash flowing seaward.

The rippled sand in the bottom of a runnel around the time of low tide shows the complexity of the local environment (figure 1.4). The mixing of the back-and-forth currents caused by waves with the shore-parallel current of the water trying to escape from the runnel forms an irregular surface. As the water flows back and forth with the waves, it creates a set of *wave ripples*—relatively symmetrical, low-relief, sometimes sharp-crested forms—nearly parallel to the shore in the bottom sediment. The alongshore flow of water rushing to escape from the runnel peaks as the tide falls. This strong, one-directional current carries sediment and creates a set of *current ripples*: asymmetrical features with the steep face downcurrent. They form across the direction of flow, so they are perpendicular to wave ripples. The result is a complex set of ripples in the bottom of the runnel. When the crossing pattern is especially clear, geologists call them *ladderback ripples*. Fossil ladderback ripples found in sandstones indicate that the sands were deposited in a ridge-and-runnel system.

Dissipative beaches, the relatively flat beaches that exist just after a storm, are at one end of an ordered classification of beach state. *Reflective beaches* are at the other end of the scale. A well-developed, reflective beach lacks a means of dissipating energy, since it has no ridge-and-runnel system or offshore bar. Reflective beaches consist of relatively coarse-grained sands and have a steep beach face. The step at the bottom of the beach face is pronounced. Instead of breaking, waves hitting a reflective beach tend to collapse and surge up the beach face. There is pronounced backwash (-swash) of both wave energy and water.

* * *

Coastal engineers have devoted a lot of effort trying to develop a mathematical formula to describe the shape of a beach profile. For decades, the expression

$$h(x) = Ax^{\frac{2}{3}}$$

has been used in an attempt to model the *"equilibrium" beach* profile. In the formula, h is the water depth at a distance; x, from the shoreline; and A, a "scale parameter" (a carefully considered fudge factor) that depends on characteristics of the sediment, usually grain size or fall velocity. As discussed in chapter 5, fall velocity (the speed with which sediment grains fall through still water) depends on the size, shape, and density of the sediment particles and on the viscosity, or thickness, of the fluid. Coastal engineers and geologists debate the formula, the concept of an equilibrium profile, and the related concept of closure depth.

In both the models and empirical observations, there clearly is a direct relationship between the size of the sediment grains and the steepness of the beach face. Beaches composed of coarse sands are steeper than those of fine sands. Gravel beaches are very steep.

Closure depth is the depth below which there is essentially no change in the beach profile over time and no sediment movement. Material such as dredge spoil placed on the bottom below closure depth is less likely to move than if it were placed in the potentially more active area above closure depth. Although the concept appears simple, a closer look reveals several complexities. One method for estimating closure depth is to plot a long time series of beach profiles on top of one another. This plot depicts the "envelope" of change. Figure 1.3F shows a simple, short-term envelope of change. When a long series of profiles is plotted, the seaward, deepest limit of variation is reasonably evident. But is the limit of profile change also the limit of sediment movement? If sediment can move across the deeper portions of the profile, does the concept of closure have meaning? Also, over how long a time period should profiles be surveyed to determine the closure depth? Because wave energy instigates sediment movement, the study should be sufficiently long to include infrequent, severe storms and their large, high-energy waves.

Given knowledge of the physics of waves and sediment movement, we can calculate the depth at which waves of given periods and heights can disturb the bottom. The depth of wave action is related to the *wave length*. During the Halloween Nor'easter or "Perfect Storm" of late October 1991, waves in the Middle Atlantic Bight had a wave *period*—the time between sequential wave crests—of 22 seconds, reached significant heights of 26 feet (8 m) well offshore, and agitated the bottom to a great depth. Using the formula

$$L = (gT^2) / 2\pi$$

where L is wave length; g, gravity; and T, wave period, a 22-second, deep-water wave would have a wave length of almost 2,500 feet (755 m) and the potential to disturb the bottom at depths of over 600 feet (200 m). In these conditions, the sediments across the entire continental shelf could have been disturbed, essentially rendering the concept of closure depth meaningless. Other calculations show that on the continental shelf of southeastern Australia, the depth of disturbance—that is, the closure depth—is greater than 100 feet (30 m) 10 percent of the time. This brings the practical application of the concept into question. Some researchers advocate selecting a theoretical closure depth based on a statistical measure of the local wave climate. One possible parameter is the wave height that is exceeded only 12 hours a year. A simpler estimate of closure depth is given by the equation

$$h_{cd} = 1.57 \, H_w$$

where h_{cd} is the closure depth and H_w, the height of the selected wave.

The calculation of closure depth is important in the planning and designing of beach-nourishment projects. If the closure depth is the seaward limit of activity, the entire region landward of closure can be active and mobile. Sediment can move back and forth between the dry beach and the active nearshore region; sand placed high on the beach should be expected to move away from shore and down toward the closure depth. Beach sand does not always stay where coastal engineers want it to. The "equilibrium" beach profile is more of an ideal concept than a reality. Its critics find fault with the underlying assumptions, and its proponents employ its utility while accepting its faults.

* * *

Beach processes are not just onshore–offshore; there also are significant longshore components. The most obvious is the *longshore current*, generated by waves approaching the beach at an angle, as they virtually always do. Waves in shallow water, especially breaking waves, transport water onto the beach. Because of the constant replenishment by the waves, the water is "pushed" ahead of the waves along the open angle between the incoming waves and the beach. Indeed, the longshore current generally is confined to the region inshore of the outermost breakers. The velocity of the longshore current depends on the size and period of the breaking waves and the angle between the waves and the shoreline. An alongshore wind blowing with the longshore current can increase the current's velocity. All of us have seen a beach ball in the water moving rapidly, sometimes very rapidly, along the shore. The beach ball travels at the speed of the wind. A better way to visualize the longshore current is to imagine watching something that floats on the water, such as a piece of driftwood, with little exposure to the wind. During class trips to the beach, coastal geologists sometimes toss a grapefruit or an orange into the water, since either fruit is easy to see, track, and retrieve (and is biodegradable, if it is lost). It is often surprising how fast the longshore current is; frequently, the current flows faster than someone can swim against it, so onlookers on the beach would see a swimmer trying to swim in one direction but actually moving backward.

The longshore current coupled with the bottom-disturbing forces of the waves can move a lot of sediment. The concentration of sediment is greatest at the bottom and decreases toward the surface. The littoral transport of sediment can be immense. It often is reported as the quantity of material passing across any particular profile line in a year. Volumes in excess of 100,000 cubic yards (76,000 m³) per year are not uncommon. This is the process by which spits grow, some inlets fill, and other inlets between barrier islands migrate.

Along a segment of a relatively straight beach, the longshore movement of sediment should not cause any noticeable change. For every quantity of sediment imported by longshore transport, a like quantity is exported down-drift. Problems, whether they be erosion of the shoreline or shoaling of channels, occur where and when something interrupts the longshore system. Problems also arise at the ends of the system, where there is no upstream source of sediment or no downstream sink.

The longshore current is not just a simple current running along the beach at a uniform velocity. Some of the variation can be a pulsing or rhythmic surging, resulting, in part, from the throbbing input of energy from waves. Surf beat and edge waves are two of these alongshore phenomena. Because both edge waves and surf beat have periods of several minutes, they are known as *infragravity waves*—that is, waves with a *frequency* below that of the regular wind or gravity waves.

Surf beat is not the audible rhythm of the breaking waves, but an oscillation of nearshore water levels caused by the interaction of two or more *wave trains* at the shoreline. These oscillations have periods of several minutes.

Edge waves are held at the shoreline by *refraction* and are caused by variations in the wave energy reaching the beach. Waves tend to arrive at the

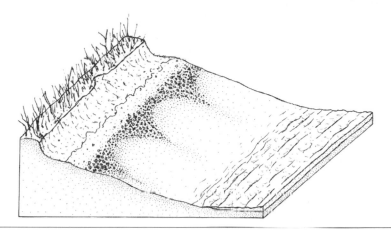

Figure 1.5 The formation of beach cusps has been attributed to either a process of self-organization or edge waves, or both. (Illustration by P. L. Mason)

shore in groups, which is why surfers wait for a good set of waves to ride. Larger waves occur in small sets separated, over some minutes, by several groups of smaller waves. This low-frequency, long-period variation in breaker heights yields a similar variability in the longshore current. In turn, this causes a low-frequency wave along the beach. In addition to a periodicity of minutes, edge waves have wave lengths that can be about half a mile (1 km or so). Their relative height decreases rapidly, exponentially, with distance from the beach. Edge waves play a significant role in nearshore hydrodynamics.

Many scientists attribute *beach cusps* to edge waves (figure 1.5). Cusps are the regular, crescentic features that sometimes develop on the outer berm and beach face of many beaches. Cusps are open (concave) toward the sea. Their spacing, or wave length, ranges from a few to tens or sometimes hundreds of feet. Cusp formation also occurs spontaneously; that is, the cusps "self-organize" as a result of interactions among the grain size of the sediments, the slope of the beach face, the angle of wave approach, and the period and height of incident wave swash. Cusps occur most often on coarse-grained beaches, including those composed of gravel and cobble, and thus are relatively steep.

Many beaches have lenses of dark-colored sand. Seen through a magnifying glass, grains of beach sand show a great variation in their color and structure. Most beach sand is composed of grains of tan or gray quartz and feldspar. Quartz (silicon dioxide [SiO_2]) is a light-colored mineral. *Feldspar*, the most abundant class of minerals, refers to a group of aluminum silicates. The majority of feldspars fall on a continuum that runs from a sodium-aluminum-silicate through a potassium-aluminum-silicate to a calcium-aluminum-silicate. Different mixtures have different mineral names. Along with the primary components, a range of other elements can substitute in the mineral structure. These variations and impurities cause the different colors of feldspar.

The dark, or black, sands can be any of a broad suite of minerals. As a group, they tend to be denser than quartz and feldspar; a handful of black sand weighs more than a handful of common, quartz–feldspar sand. Indeed, most of the dark sands fall into the category of *heavy minerals*, which have densities, or specific gravities, greater than 2.89. *Specific gravity* is the relative density of a substance compared with the density of water. The density of quartz is 2.65, and the various feldspars cluster fairly tightly around that. Magnetite, an iron oxide commonly found as black sand, has a density of 5.18. Garnet, another common component of beach sands in New England, has a density of at least 3.5. Black sands usually are finer grained than quartz–feldspar sands. It is the difference in density and grain size that allows for the natural separation of the heavy, dark sands from the lighter matrix.

Transport in water is a mechanism for sorting sediments, which occurs as a function of the size, density, and shape of the particles. These factors are mainly responsible for determining how the particles act in water, including how fast they settle to the bottom. Generally, smaller but denser particles settle at the same speed as larger, lighter particles—hence the usual mixture of beach sediments. But when sediments are subjected to energetic conditions over a relatively long period, separation occurs. It is not just that heavy minerals settle more rapidly than light ones, but that sometimes they are left behind when the coarser though less dense sediments are eroded. The process is roughly analogous to panning for gold; as the prospector swishes the water around the pan, the dense flakes of gold stay while the lighter particles are carried away. Settling can be a self-reinforcing process; once there is a concentration of heavy-mineral sands, that concentration tends to remain and grow.

In some locations with a concentration of specific heavy minerals that have economic value, beach sands have been actively mined. Some sites are "fossil" beaches, ancient shorelines that no longer are active. Beach sands in Alaska have been mined for gold. Trail Ridge, a fossil beach in northeastern Florida, has been mined for the titanium minerals ilmenite, rutile, and leucoxene and for other ores for nearly a century.

* * *

We usually do not think of beaches as habitat, as places where creatures live. When at the shore, we notice the other beachgoers enjoying the surf and the seagulls wheeling around in the sky, alert for tidbits to scavenge. But there is other life on and in the beach. Ghost crabs (*Ocypode* species) are important beach dwellers. They are significant predators and scavengers who excavate burrows that can be as much as 3 feet (1 m) long. Many coastal scientists use the status of a beach's ghost-crab population as an indicator of the environmental health of the beach. As they zigzag across the beach at speeds up to 6 feet (2 m) per second, ghost crabs provide entertainment for us, especially when our pet dog tries to catch them.

Sea turtles crawl out of the ocean and onto the beach to lay their eggs. Birds like the Least Tern (*Sternula antillarum*) and the Piping Plover (*Charadrius melodus*) nest on the beach. Designated a "threatened species," the Piping Plover nests on the back of the berm near the sand dunes. Resource managers close beaches to use by vehicles to protect nesting Piping Plovers. The Least Tern uses shallow "scrapes" on the beach and in other sandy or gravelly settings.

In *The Edge of the Sea*, Rachel Carson described smaller animals that live in the beach sand. Sand fleas, any of several small pests that live on the beach, are not insects but are crustaceans that, like their insect namesakes, annoyingly bite our ankles. Most of the beach-dwelling organisms are quite small, often less than 0.04 inch (1 mm) in length. If you think about the harsh physical environment in which they live, you soon realize that they have to be pretty tough, with hard outer-body parts to protect them from being abraded and crushed. Although it may be disquieting to think of all the "bugs" living in our sandy playground, it is good that they are there. They help make the beach a better place by cleaning up the rotting matter that would stink and foul the otherwise apparently clean environment. Even bacteria are important in beach processes. For example, a particular bacterium helps calcium carbonate (calcite [$CaCO_3$]) precipitate out of seawater and bind sand grains together to form beach rock.

* * *

Beaches are complex and dynamic settings. In order to understand what a beach is and how and why it changes, we have to think about the forces that work in concert to shape both beaches and the fragile areas that adjoin them.

2
WIND

An afternoon breeze off the water lessens the discomfort of a hot summer day in the sun, flies kites, dries bathing suits and beach towels, and blows gnats and mosquitoes away from the beach. Wind, especially storm wind, is an important part of beach processes. Very strong winds, such as hurricanes, affect much more than just the beach. Even moderate winds generate waves, which are important mechanisms that alter the shore and which add to our enjoyment of a day at the beach.

The sun is the essential driver of wind. As the sun heats the atmosphere, the sea, and the land, they warm unevenly. During a sunny summer day, the top few inches of dark soil and grass absorb a lot of solar energy (heat). Light-colored areas—such as sandy beaches, snow, and the polar ice caps—reflect much of the incoming energy back through the atmosphere and into outer space. The ocean both absorbs and reflects the solar energy. The term for this reflection is *albedo*. The higher the percentage of incoming, or incident, solar energy that a surface reflects, the higher the albedo of that surface.

As solar energy warms the sea surface, waves and other processes mix the water. Anyone who has swum in a New England lake or mountain pond knows that the sun heats the upper few inches, while the water below the surface remains quite chilly. Because there are essentially no waves on a pond, the warmer surface water does not mix down into the colder subsurface water, so someone diving into the pond often finds uncomfortably cold water. Such a sharp change in temperature is called a *thermocline*. Because of the mixing of surface and subsurface water by waves, there rarely is a thermocline close to the surface near the shore of the ocean or a large lake.

Wind is air in motion. The air itself is heated by both the incoming sunlight and the heat radiated from Earth. The air in a region is not heated uniformly, often because Earth's surface loses heat at varying rates. Characteristics like the color of the surface and the nature of the material influence how much heat is reflected and how fast heat is released, or radiated back, into the atmosphere. As a mixture of gases, air expands when heated and contracts when cooled. Therefore, air expands or contracts to different degrees in different places, at different times, all at different scales of size and time. A barometer measures the air pressure, which indirectly measures the density of the air. Like any other fluid, air flows from places of high pressure to places of low pressure. The wind is stronger when the difference in barometric pressure between two places is greater. Put another way, a large pressure differential (the difference in air pressure divided by the distance between the two places where it is measured) causes a high wind.

Speed, direction, and duration are the properties that determine the importance of the wind in shore processes. The impact of speed is easy to understand, and, indeed, our day-to-day terminology tells the story. "Strong" winds have fast speeds. Higher winds create higher waves and can push the water to a higher *storm tide*.

On a practical level, the speed or strength of the wind determines how we respond to it. The television weather forecaster is good at telling viewers to bring in lightweight lawn furniture when the wind speed is predicted to rise or to tie down the picnic table if the wind is really going to blow. Calculating the actual force of the wind is fairly simple. It is a basic physical principle that the force of a moving object is equal to the mass (weight) of the object times the square of its velocity (speed). Many of us learned this rule in a high-school science class as the equation

$$F = mv^2$$

where F is the force applied by something (in this case, the wind); m, that thing's mass; and v, the velocity at which that mass is moving. Even though we seldom think of it, air does have weight; under average conditions, 1 cubic foot (0.03 m³) of air weighs about 1.3 ounces (37 g). Double the wind speed, and quadruple the force it applies; double the wind speed again, and increase the force by another factor of four. Thus a wind blowing at 20 miles per hour (mph) (32 km/hr) has four times the force—exerts four times as much pressure on a tree, window, or wall—as a 10-mph wind. A 40-mph wind pushes 16 times as hard as a 10-mph wind, and a storm with 80-mph winds exerts 64 times the pressure on whatever it blows against as a 10-mph breeze. Consider the numbers for v^2:

v	v^2
10	100
20	400
40	1,600
80	6,400

No matter what units of measure you use—miles per hour, knots, meters per second—the same ratios hold true. Another important reminder is that "knot" is the term for a unit of speed; 1 knot is 1 nautical mile (about 1.15 statute miles) per hour. People who talk about wind speed or their boat's cruising speed as so many "knots per hour" are speaking incorrectly. (For a table of conversions among the various measurement systems, see appendix 1.)

Another commonly used method of reporting wind speed is the Beaufort Scale (appendix 2), which Admiral Sir Francis Beaufort developed in the early nineteenth century. By 1838, the British Royal Navy required the use of the Beaufort Scale for reporting wind velocity. It appears in maritime weather broadcasts in many parts of the world. Originally, the scale had 13 categories numbered from 0 (flat calm) to 12 (hurricane). Later, categories 13 through 17 were added to encompass stronger hurricanes.

The title of Alistair MacLean's thriller *Force 10 from Navarone* (1968) describes the storm that plays an important part in the novel's action.

Interestingly, although the Beaufort Scale relates to the speed of the wind, it really assesses its force. Whitecaps first appear when the wind is at Force 3, about 10 knots. When the wind is strong enough to break twigs and small branches from trees, it is at Force 8 (gale), about 34 to 40 knots. This is the practical advantage of the Beaufort Scale: no instrument or tool is needed to measure the wind speed; you only have to observe the effect of the (force of the) wind on the world around you. Scott Hurler's book *Defining the Wind* provides more information about Beaufort, the history of wind observations, and the Beaufort Scale.

More recently, weather forecasters have come to use the Saffir-Simpson Hurricane Scale (appendix 3) to rate hurricanes from 1 to 5, depending on wind speed. The scale describes the destructive capabilities of the storms. Much like the Beaufort Scale, the Saffir-Simpson Scale relates the effective force of the wind to its velocity. Sometimes it helps to view the force of the wind from a human perspective. While the wind of a Category 1 hurricane might blow you off your feet and push you backward as you try to walk into it, the wind of a moderate Category 3 hurricane is strong enough to pick you off the ground and carry you some distance. That wind speed of 125 miles per hour (200 km/hr) is the same as the maximum speed that sky divers attain when free-falling before opening their parachutes.

As important as the wind is in defining the strength of hurricanes or other storms, however, most of the damage and loss of life associated with major coastal storms comes from flooding. Both the storm-driven rise of water level, known as storm surge, and the inundation due to the very heavy rain are the major agents of destruction. Only in the descriptions of the higher Saffir-Simpson categories does wind-driven damage become important.

The weight of the air also has a role in determining the force of the wind. Air density, the weight of a set volume of air, is related to the m (mass) in the equation $F = mv^2$. Air density decreases with increasing humidity, temperature, and altitude (but since beaches are at sea level, altitude does not affect our thoughts there). Because a cold wind is denser than a warm one, it has more force than a warm wind of the same speed. One cubic foot (0.03 m³) of cold air weighs more (is denser) than the same volume of warm air. Thus during the winter, whitecaps begin to form at slightly lower wind speeds than during the summer because warm wind needs a slightly greater velocity to reach the same force.

Surprisingly, dry air is denser than wet air. Although we often refer to humid summer air as "heavy," airplane pilots will tell you that flight is more difficult in humid air because it is less dense than dry air. The lower density of humid air occurs because the molecular weight of water, approximately 18, is less than the molecular weight of dry air, which is approximately 29.

Going back to basic chemistry, the symbol for water is H_2O (two hydrogen atoms and one oxygen atom), with a hydrogen atom having an approximate atomic weight of 1 and oxygen an atomic weight of 16 (so water has an atomic weight of 18). But the atmosphere is mostly nitrogen (N_2), with a nitrogen atom having an atomic weight of 14 (so the two atoms comprising the N_2 molecule have an atomic weight of 28). Thus dry air, which is about 78 percent nitrogen and 21 percent oxygen, is denser than humid air because increasing the water content decreases the molecular weight of the air. Therefore, an airplane has to work harder to fly on a hot, humid day than on a cold, dry day because the lower-density air provides less lift.

The direction of the wind relative to the beach affects the beach. The three basic wind directions are onshore, offshore, and alongshore. Wind is described by the direction from which it blows: a north wind blows from north to south, and a sea breeze blows from the water toward the land. A wind described as "335 degrees at 17, gusting to 25," comes from 335 degrees (true)—that is, from the northwest—and has a steady speed of 17, with occasional peaks of 25. For maritime or nautical weather reports in the United States, we should assume that the "unit" for maritime wind, if not specified, is knots.

The duration of the wind sometimes is related to the physical size of the wind system. Local sea breezes last for only a few hours. *Tropical storms*, hurricanes, and *extratropical storms*—which we sometimes call nor'easters—may continue for many days but usually affect a particular locality for one day. However, some nor'easters do have an impact on local areas for a few days. Finally, winds related to global circulation patterns, such as the trade winds, blow almost continuously.

The local sea breeze is a great example of how the sun causes wind. Anyone who has spent a summer at the shore has experienced the near-daily phenomenon in which a breeze blowing from the water onto the land develops through the afternoon. The sea breeze occurs on a sunny summer day because the surface of the land heats up faster than the surface of the water. As the day progresses, the relatively hot land radiates heat back into the atmosphere, warming the air immediately over the land. The warmed air expands, becomes less dense, and thus rises. The denser, cooler air over the water flows toward the land to replace the rising heated air. As sailors know, the strength of the sea breeze reaches a maximum in the late afternoon. This is when the temperature differential is greatest. As the evening gathers, the land surface continues to radiate heat, but the wind diminishes as the difference in temperature between the land and the sea decreases. Eventually, when the land becomes cooler than the water, the wind direction reverses.

In addition to generating waves, wind moves water. Storm surge and *Ekman transport* are two of the major ways in which wind forces water. During the Norwegian North Polar Expedition of the mid-1890s, the oceanographer Fridtjof Nansen observed that sea ice moved along a course 20 to 40 degrees to the right of the wind. He theorized that the deviation between the direction of water movement and that of the wind was due, in large part, to Earth's rotation and that the deviation increased with depth. Later, the Swedish oceanographer Vagn Walfrid Ekman mathematically verified Nansen's hypothesis.

These studies resulted in a concept of wind-driven water movement that has come to be known as the *Ekman spiral*, which is defined in the *Glossary of Oceanographic Terms*:

> A theoretical representation of the effect that a wind blowing steadily over an ocean of unlimited depth and extent and of uniform viscosity would cause the surface layer to drift at an angle of 45 degrees to the right

of the wind direction in the Northern Hemisphere. Water at successive depths would drift in directions more to the right until at some depth it would move in the direction opposite to the wind. Velocity decreases with depth throughout the spiral. The depth at which this reversal occurs is on the order of 100 meters [300 feet]. The net water transport is 90 degrees to right of the direction of the wind in the Northern Hemisphere.

In the Southern Hemisphere, the shift is to the left. Real conditions differ from theory, as theory requires the nonexistent "ocean of unlimited depth and extent and of uniform viscosity."

In the real world, the motion of the sea ice is less than 45 degrees off the wind because the ice is pushed by both the wind itself and the wind-generated current in the water. The "sail area" of the ice exposed to the wind, the strength of the wind, the keel of the iceberg exposed to the water, the draft of the ice in the water, and the strength of the current act together to determine the course of the ice.

Upwelling is an important consequence of the wind-generated water motion. Consider the example of a prolonged south wind blowing along the East Coast of the United States during the summer. Ekman transport causes the sun-warmed, surface water to flow offshore toward the northeast. Cooler water from deeper in the ocean wells up to replace the water moved northeast by the wind (figure 2.1). This process has spoiled the vacations of countless people who have tried to escape the hot, south wind by going to the beach, only to find the water uncomfortably cool for swimming. But improved fishing sometimes compensates for cold swimming. The cooler, upwelling water can contain more dissolved oxygen than the warmer water it displaces. The oxygen-rich water can support more fish. Thus upwelling can simultaneously decrease swimming comfort, because the water is chilly, and improve the local recreational and commercial fisheries. Such upwelling contributes to the robust commercial fishery off the Pacific coast of South America. Strong south winds move the surface water offshore (leftward because of the continent's location in the Southern Hemisphere). The coast of Ecuador, Peru, and Chile has an extremely narrow continental shelf before the seafloor drops into

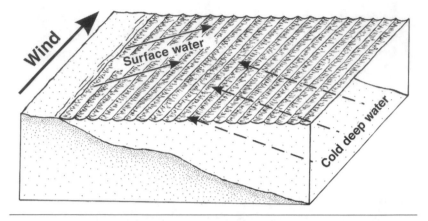

Figure 2.1 Upwelling: because of Ekman transport, the surface water moves offshore in response to wind blowing parallel to shore. Colder water from deeper offshore areas flows up to replace the surface water. (Illustration by P. L. Mason)

very deep water. Therefore, the water that upwells is cold and nutrient rich. These nutrients form the base of the food chain that supports the healthy fishery.

Ekman transport also influences the impact that a storm may have on the shore. During the nor'easters that are common along the Middle Atlantic and New England coasts of the United States, the sustained northeast wind causes Ekman transport that flows from the east to the west. This contributes to storm surges, especially in bodies of water that are open to the east, such as Chesapeake, Delaware, and New York bays.

* * *

Wind affects the shore in several ways. It generates the waves that shape the beach. Its strength determines the classification of hurricanes and other storms, whose winds damage structures and cause trees to sway and break. Wind can move water, both toward the shore as a storm surge and away from the shore. With Ekman transport, it leads to the upwelling of colder water. Finally, as we will see, wind moves sediment and forms sand dunes.

3
WAVES

Waves move energy from one place to another. Waves that break on the beach and rock boats also erode the shore and modify beaches. Wind causes most ocean waves, so they are called *wind waves*. When waves are in deep water and cannot interact with the seafloor, gravity is the major force acting to damp them; gravity tries to restore the water to a flat surface. Because of the importance of gravity in the physics of wind waves, they also are called *gravity waves*. There are several other ways to classify waves.

When wind, even a light breeze, blows across a flat water surface, there is friction between the moving air and the still water. The resulting drag causes ripples. Rippled water has both a greater surface area than flat water and a vertical relief against which the wind can press. This results in more efficient transfer of energy from wind to water, so the ripples grow into waves fairly rapidly. In deep water, three factors work together to determine and limit the size of the waves: the strength (or speed) of the wind; the length of time it blows; and the unobstructed, overwater distance across which it blows. The term for the last factor is *fetch*. Increasing any one of the three factors allows the size of the waves to increase unless there is some limiting influence. Some oceanographic texts and manuals have tables that relate wind speed, wind duration, and fetch to the size of the waves. In shallow water, depth also plays a role.

To talk about waves, oceanographers have developed a standard vocabulary, and they use a standard set of symbols to represent the specific terms (figure 3.1). The shape of a wave, really the shape of a cross section of the wave that parallels the direction in which the wave is moving, is a *trochoid* (figure 3.2). You can draw a trochoid by tracing the curve that one point on the radius of a circle travels as the circle rolls across a flat surface.

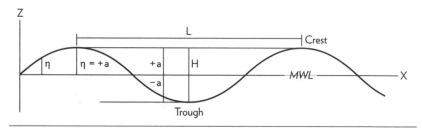

Figure 3.1 Wave terminology: MLW, mean water level (sometimes called still water level); L, wave length (horizontal distance from crest to crest or trough to trough [sometimes symbolized by λ (lambda)]); H, wave height (vertical distance between crest and trough); a, amplitude (½ H, or vertical distance from MWL to crest or trough); η (eta), vertical distance from MLW to water surface; X, horizontal dimension (positive to the right); Z, vertical dimension (positive above MLW, negative below MLW); crest, high point of the wave; trough, low point of the wave; h (not shown), water depth. (Illustration redrawn by Network Graphics from an original by the author)

Figure 3.2 A trochoid wave: before breaking, a wave can reach a maximum steepness of 120 degrees and 1:7 wave height:wave length ratio. A trochoid shape can be made by following a point on the radius of a circle as the circle rolls against a flat surface. To obtain the shape of an ocean wave, the flat surface has to be on top of the circle, or, if the surface is beneath the circle, flip the resulting curve. (Illustration redrawn by Network Graphics from an original by the author)

The energy of the wave moves with the shape of the wave, but there is relatively little net movement of the water. This is easy to see by watching a piece of *flotsam* or the float on a fishing line as it rides on gentle waves. It moves forward with the wave crest and returns toward its original position with the trough. The wave shape and wave energy continue to move forward, but the float and the water cycle back and forth. Any net motion of the float likely is the result of an underlying current or the wind.

Within a wave, the water can be thought of as moving in a (vertical) circle as the wave passes. At the crest of the wave, a particle of water moves in the same direction as the wave; in the trough, the particle moves back; and at the midpoints, the particle moves vertically, up or down, depending on which side of the circle is being observed. All this water motion has an impact beneath the surface. The motion continues, but decreasingly so, with depth. The circle becomes smaller and smaller as the depth increases, but the time it takes the circle to make a full revolution remains the same; thus at depth, the rim of the circle covers less distance during the same time, so its speed is less. Just as with wind, lower speed means lower force; therefore, the deeper something—the sea bottom, a fish, or a submarine—is below the surface, the less force a wave can exert on it. The reduction in force is predictable, as the diameter of the circle decreases by one-half with an increase in depth of one-ninth of the wave length (table 3.1).

Shorter diameters of wave motion are fine in deep water, but when the water is so shallow that the circles reach or "feel" the bottom, things change. Still thinking of the circle analogy, in shallow water the downward arc "hits" the bottom, cannot continue its downward motion, and turns to move horizontally across the floor in the same direction as the unhindered wave. The water moving across the bottom is subject to friction. Thus as the circle flattens, the water motion tends to be bidirectional, back and forth, as opposed to the original circular pattern. When we are in the water at the beach, we feel these dual motions. We are pushed landward by the wave (crest) and drift seaward in the trough. The extreme example of this is on the beach face with the uprush and backwash of the waves.

TABLE 3.1

Decrease in Water Motion with Depth

Depth in wave lengths (L)	Diameter (relative)
Surface	1
$\frac{1}{9}$	$\frac{1}{2}$
$\frac{2}{9}$	$\frac{1}{4}$
$\frac{3}{9}$ ($\frac{1}{3}$)	$\frac{1}{8}$
$\frac{4}{9}$	$\frac{1}{16}$
$\frac{5}{9}$	$\frac{1}{32}$
$\frac{6}{9}$ ($\frac{2}{3}$)	$\frac{1}{64}$
$\frac{7}{9}$	$\frac{1}{256}$
$\frac{8}{9}$	$\frac{1}{256}$
$\frac{9}{9}$ (1 wave length)	$\frac{1}{512}$

Note: The ability of a wave to disturb the sediments on the sea floor is related to the wave length and the water depth. Water moves in a circular pattern beneath a wave. The diameter of the circle decreases by one-half with an increase in depth of one-ninth of the wave length. As the wave period does not change with depth, the speed of water movement—hence its ability to move sediment—beneath a wave also decreases with depth.

Storm waves disturb the bottom at a greater depth than do smaller waves because of the relationship between wave length and wave motion at depth. Storm waves have relatively long lengths, so the wave motion reaches deeper. Or if the water is shallower, the strength of the wave motion on the bottom is greater than that of the shorter, fair-weather waves. When the motion across the bottom is strong enough to move the sediment on the seafloor, turbulent water can lift the sediment and carry it elsewhere. *Wave base* is the maximum depth at which waves can initiate movement of bottom sediment. Wave base influences the offshore disposal of dredge spoil. Although the concept is fairly simple, it is often difficult to agree on which wave conditions to use when calculating wave base for an engineering purpose. Should engineers make their calculations for the largest waves possible at the location; for waves that are exceeded only, say, 1 percent of the time during any 10-year period; or for a lesser, but more frequent standard?

TABLE 3.2

Deep-Water, Transition, and Shallow-Water Waves

Deep-water wave	$h > \frac{1}{2} L$	$h > 0.5\,L$
Transition wave	$\frac{1}{2}\,L > h > \frac{1}{20}\,L$	$0.5\,L > h > 0.05\,L$
Shallow-water wave	$h < \frac{1}{20}\,L$	$h < 0.05\,L$

where h is the depth of the water and L, the wave length.

Clearly, waves in shallow water "feel" the bottom a lot more than do waves in deep water. For this reason, we distinguish deep-water waves, transition waves, and shallow-water waves (table 3.2). A wave is a *deep-water wave* if it is in water with a depth greater than 50 percent (one-half) of its length. A wave is a *shallow-water wave* if it is in water with a depth less than 5 percent (one-twentieth) of its length. Transition waves occur between deep- and shallow-water waves. Waves change significantly as they move from deep to shallow water.

Because waves move, it also is important to know their speed. C is the symbol for the speed at which a wave moves. It stands for "celerity," from the Latin term *celeritas* (speed); the word "accelerate" comes from the same root. Another basic measure of a wave is its period, noted with the letter T. Wave period is the time it takes one full wave length to pass a fixed point. Physical oceanographers often describe waves by their period—for example, a 6-second wave. Longer-period waves are larger and more powerful than shorter-period waves. Sometimes it is easier to use frequency instead of period. A wave's frequency, denoted by the letter F, is the number of wave lengths that pass a fixed point during a standard time interval, such as 10 waves per minute. Frequency and period are the inverses of each other (box 3.1).

Now we can begin to combine the characteristics of waves. It is important to understand the general ways at which wave length, the speed with which the wave moves, water depth, and the height of the wave are interrelated. Although the relationships are mathematical, I will try to describe them in words and separate the equations. Do not be put off by the math, for it simplifies pretty quickly.

Frequency (*F*) is 1 divided by the wave period (*T*):

$$F = 1 / T$$

Wave period is 1 divided by the frequency:

$$T = 1 / F$$

Speed (*C*), whether that of a wave in the ocean or a car on the highway, is the distance (*L*) traveled in a given time (*T*), which is calculated as distance divided by time. We have to know, or to measure, two of the three variables to obtain the third. There is, however, an equation that relates *C* and *L* in waves (box 3.2). The equation says that the speed of the wave is related to the wave length (*L*), the water depth (*h*), and gravity (*g*). Gravity is a constant: 32 feet per second squared (980 cm per second squared), which is expressed as 32 ft/s^2 (980 cm/s^2). Figure 3.3 shows the speed of a shallow-water wave in different water depths.

The equations in box 3.2 explain a rule of thumb that many boaters use to estimate the maximum speed of a displacement-hull (as opposed to a planing-hull) craft. The rule states that the hull speed, in knots, of a boat is 1.3 times the square root of the waterline length in feet:

$$C = 1.3(\sqrt{L})$$

where *L* is the length of the boat's waterline. The fastest that a displacement-hull vessel can travel is the speed of a wave equal to the waterline length of the vessel. If the vessel went faster, it would have to climb the wave and would plane.

Using this relationship, a displacement-hull boat with a 25-foot (8-m) waterline length has a maximum speed of about 1.3 × √25 (1.3 × 5), or 6.5 knots; a vessel with a 49-foot (15-m) waterline length has a maximum speed of 1.3 × √49 (1.3 × 7), or 9.1 knots; and a boat with a 100-foot (30-m) waterline length has a maximum speed of 1.3 × √100 (1.3 × 10), or

BOX 3.2

Speed (*C*) is distance (*L*) divided by time (*T*):

$$C = L \,/\, T$$

or

$$C = L \div T$$

The equation for the speed of a wave is (don't panic, it will be simplified)

$$C = \sqrt{\frac{gL}{2\Pi} \tanh \frac{2\Pi h}{L}}$$

where *g* is gravity and *h*, water depth. The inclusion of the term π (pi) suggests a circle and refers to the discussion of the circular movement of water in a wave. Tanh (the hyperbolic tangent) is a term from trigonometry.

The equations for deep- and shallow-water waves can be simplified. In deep-water waves, *h* / *L* is fairly large, >0.5, so tanh(2π*h* / *L*) approaches 1 and can be ignored because multiplying a quantity by 1 does not change it. For the minimum deep-water wave (that is, a wave in water one-half as deep as the wave is long), tanh(2π*h* / *L*) = 0.996; for a wave in water as deep as the wave is long, tanh(2π*h* / *L*) = 0.99999. As the values for these hyperbolic tangents are really close to 1, we reduce the equation for deep-water waves to

$$C = \sqrt{(gL \,/\, 2\pi)}$$

or the square root of the acceleration of gravity times wave length divided by 2 pi.

The simplification for shallow-water waves is similar. Here, tanh(2π*h* / *L*) approaches the value 2π*h* / *L*, simplifying the equation to

$$C = \sqrt{[(gL \,/\, 2\pi)(2\pi h \,/\, L)]}$$

which, because the *L*s and 2πs cancel, further reduces to

$$C = \sqrt{(gh)}$$

or the square root of gravity times depth.

There is no similar simplification for transition waves; so rather than struggle through the full wave equation and because transition waves occur in only a limited area and their properties are intermediate between deep- and shallow-water waves, we will not consider them further.

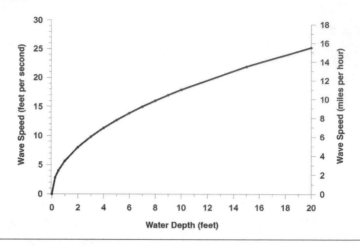

Figure 3.3 A plot of water depth versus the speed of a shallow-water wave using the formula $C = \sqrt{(gh)}$.

approximately 13 knots (box 3.3). Again, this is a rule of thumb that works pretty well for recreational-class vessels. As boats increase in length, beam, and displacement, other factors come into play. As boats with a planing hull can be thought of as riding on top of the water, they can go substantially faster than those with a displacement hull, and the equation does not apply.

The wave equations in boxes 3.1 and 3.2 explain some phenomena that beachgoers observe at the shore. Comparing the simplified equations for shallow- and deep-water waves confirms wave behavior that seems intuitive: as the water gets shallower, the wave speed decreases. The increasing friction from the wave "feeling" the bottom slows the wave.

Given the two equations

$$C = \sqrt{(gL / 2\pi)} \text{ and } C = \sqrt{(gh)}$$

look at the relationship of $L / 2\pi$ and h. In a shallow-water wave, h is less than $0.05L$, or, stated another way, $h < L / 20$, and as the equation calls for $L / 2\pi$, or $L / 6.28$, C in deep water has to be greater than in shallow water. Wave length and water depth have important consequences in how waves behave.

BOX 3.3

To see how the rule of thumb for the maximum hull speed of a displacement-hull boat was derived, look at the simplified equation for a deep-water wave:

$$C = \sqrt{(gL / 2\pi)}$$

Squaring that equation yields

$$C^2 = gL / 2\pi$$

Here, $g / 2\pi$ is a constant, $(32 \text{ ft/s}^2) / (2 \times 3.14)$, which, doing the division, equals 5.09 ft/s^2, with s being seconds. So

$$C^2 = L \text{ ft} \times 5.09 \text{ ft/s}^2$$

Taking the square root of both sides of the equation,

$$C = \sqrt{[(5.09L \text{ ft}^2)/s^2]}$$

which reduces to

$$C = (2.26 \text{ ft}\sqrt{L}) / s$$

We know that 1 knot equals 1.7 feet per second, so

$$2.26 / 1.7 = 1.3$$

Thus

$$C \text{ (in knots)} = 1.3\sqrt{L}$$

with L being the waterline length in feet.

Think of what happens as a wave slows but the period, the time between waves, remains unchanged. Because speed equals distance divided by time ($C = L / T$), if the speed decreases and the time interval is constant, length also has to decrease. So as waves approach shore, the wave length becomes shorter. In effect, waves try to catch up to the preceding waves. Think of a line of absolutely identical race cars that are evenly spaced 100 feet (30 m) apart and going 120 miles per hour (193 km/hr) on a long straight section of a racetrack. By doing the math, the cars are 0.57 second apart. When they slow down to go into a corner, the cars get closer together, but remain the same time interval apart. If they slow to 60 mph

(97 km/hr), the distance between them will shorten to 50 feet (15 m), but the time between them remains 0.57 second. Similarly, the wave crests get closer together in distance, although the wave period remains the same. But the quantity of water in motion does not change. Because the ability of the water in the wave to move down is limited by the shallow bottom, the only direction the water can move is up. The waves get taller; H increases. This process works until the waves become too steep to hold together.

Wave steepness is the ratio of height to length (H:L). The maximum attainable wave steepness is 1:7. Above that, the wave becomes unstable and begins to break, or fall apart, because water as a fluid does not have enough "internal strength" to hold together. A similar limiting factor is the wave angle centered on the wave crest between the front and back faces of the wave. At wave angles sharper than 120 degrees, the wave becomes unstable and breaks (see figure 3.2). The instability also occurs because the forward motion of the water in the upper part of the wave is faster than the advance of the wave shape. Waves break in water that is about one-third deeper than the wave is high.

If our observation of waves changes from a cross section to a map view, such as when looking straight down from an airplane, we would see the clear effect of waves slowing as the water becomes shallower. In a simple case, a straight shoreline has a seafloor with a constant slope toward the offshore and deep-water waves that approach the coast directly. As the wave crests arrive parallel to the beach, they become closer together (figure 3.4). More often, though, the deep-water waves approach the shore at an angle. Individual waves slow at different rates along the wave crest, making the wave crests become more parallel to the beach as they approach it (figure 3.5). This phenomenon of "bending" is called wave refraction. The refraction of water waves is similar to the bending of light waves in eyeglasses and telescopes. Snell's Law, which characterizes the refraction of both water and light waves, relates the amount of bending to the speed of the wave and some (changing) characteristic of the medium through which the wave moves.

Currents also can change the apparent speed of waves. The formulas yield the wave speed in the ambient (surrounding) water. If the water has a velocity of its own—for example, a tidal current or the outflow of a river—the wave speed "over the bottom" is the sum of the wave and current speeds. Think of a straight shoreline with a uniform offshore slope and the deep-water wave crests parallel to shore. If there are no other influences, simple wave refraction keeps the waves parallel to one another and to the beach; the waves slow and become closer together as they near

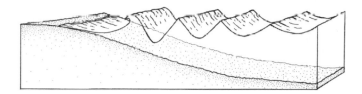

Figure 3.4 Waves slow when they move into shallower water. When waves approach a gently sloping shore straight on, they get closer together (the wave length shortens), and wave height and steepness increase. The wave period (the time between the passage of successive wave crests at a fixed point) remains constant. (Illustration by P. L. Mason)

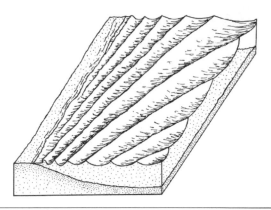

Figure 3.5 Wave refraction occurs where waves approach a gently sloping shore at an angle. Because the waves slow as they progress through shallowing water, the leading sections of the waves slow before the trailing portions, decreasing the distance between the wave crests. Thus the waves tend to become roughly parallel to the shore instead of remaining at the original angle of approach. (Illustration by P. L. Mason)

the shore. But if in the center of the beach there is a river mouth with a current flowing straight out, perpendicular to the shore, the incoming, deep-water waves encounter the current. While keeping the same "water speed," the waves lose speed "over the bottom." The slowed portion of the waves steepen and move closer together farther offshore than does the portion of the waves that are not affected by the current.

Local changes in the shape of the bottom near a river mouth or tidal inlet further complicate the situation. Shoals or underwater portions of deltas are areas of shallow water farther offshore, and channels cut through these features are areas of locally deep water. The rapid changes in topography interact with the currents near an inlet to yield sharp alterations in wave characteristics that, for instance, make handling a boat difficult.

* * *

So far this discussion has considered only waves moving in one direction and, usually, only uniform waves. Nature rarely is so simple. When strong winds blow at sea for a long time over deep water, waves with a wide range of heights, lengths, and periods exist at the same time. Because long waves travel faster than short waves, they arrive at the shore before short waves generated at the same time and place. Some waves are just forming, some have been growing for a while, and some date from the onset of the wind. Indeed, new waves can be forming on the backs of older waves. Hence you observe the confused ocean surface with waves of different sizes moving in different directions. This actually is termed a *sea*. When the wind stops or when the waves travel away from the wind that formed them, the sea tends to even out as the longer waves move away more rapidly, leaving the shorter waves behind. This yields the calmer surface and relatively uniform *swell* with which most beachgoers and seafarers are familiar. If the long waves were created by a storm, they may hit the shore well before the storm itself. The large swell at the beach in Galveston, Texas, the day before the great hurricane of September 8, 1900, should have been a warning. Before our era of very rapid communication, Pacific Islanders "knew" when a typhoon was coming because they were aware of unexpectedly large waves crashing on their shores.

Two other processes can change the direction in which waves move: reflection and diffraction. Reflection is just what it says. When a wave strikes a rigid vertical surface, it bounces off and heads back to sea. Just like a billiard ball striking the bumper or a flashlight beam hitting a mirror, the angle of incidence equals the angle of reflection. The incoming, or incident, and the outgoing reflected waves interact in a predictable manner: they "add" together. Where the crests of the incident and reflected waves coincide, the joint wave is quite large. Where the troughs coincide, the combined trough is deep. Where a crest and a trough coincide, the surface approaches an average elevation. Local conditions in the region of wave reflection can be much more severe than the prevailing conditions a little farther offshore. Also, the currents beneath the superimposed waves tend to be relatively strong and to reverse direction abruptly. The most extreme example is where the waves approach straight on and the combined incident and reflected waves can form a standing, or stationary, wave in front of a seawall. This type of interaction is called *clapotis*.

Reflection also is an important reason that sand rarely remains in front of seawalls and bulkheads. The energy of the combined incident and reflected waves lifts the sediment off the seafloor so that it moves with local currents and does not remain in front of the reflecting wall.

Wave *diffraction* occurs when wave energy spreads parallel to the wave crest as a wave passes a steep-sided obstruction. Diffraction occurs in water of any depth. One way to visualize diffraction is to think of waves moving from one direction toward a spit of land that rises steeply from the seafloor. The area immediately in the lee of the spit is not quite in a "shadow" from the waves, so there is some reduced wave action in the lee area (figure 3.6). The process by which waves propagate, or disburse, into the shadow is diffraction. Marine and coastal engineers who design harbors, breakwaters, and jetties must be aware of diffraction and of an area's wave climate so that their anchorages and harbors are safe and do not experience unexpectedly rough water during storms.

As on the open ocean subjected to strong winds and as with wave reflection, multiple wave trains often interact. Wave trains can cross one another at angles; old, long-period swell can overtake and interact with younger shorter swell; and so on. Usually one set of waves appears larger

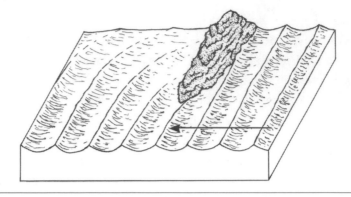

Figure 3.6 Wave diffraction occurs where waves passing an obstruction naturally extend into the shadow zone of undisturbed water in the lee of the obstruction. (Illustration by P. L. Mason)

or dominant, but the prudent mariner must be aware of the crossing swells, especially as the interaction tends to be constructive or cumulative; that is, the crests combine to create occasional extra-high waves. This piling-up of wave crests explains some of the "rogue waves" that occasionally make news by damaging ships at sea.

As longer, faster waves overtake shorter, slower waves, and as waves move into otherwise still water, other phenomena occur. It only makes sense that waves have to expend some of their energy as they roll into previously calm water. In 1844, J. Scott Russell, a Scottish naval architect and engineer who pioneered studies of solitary waves, was working with a wave-generating machine in a long channel; he noted that when a wave train advances into calm water, the leading wave decays and a new wave forms at the rear of the group. However, the new wave may not be as distinct or as well formed as its predecessors. The quantity of energy in the wave train remains (very nearly) constant. But while the speed of any individual wave does not change, the speed of the wave train slows considerably. This makes sense when you think of timing a wave train as it moves between two points. You start your stopwatch when the first wave reaches the first point and stop it when the leading edge of the group reaches the second point. Remember that the first wave decays, so the "new" first wave has had farther to travel because it was back in the group when the original first wave passed the starting point. The speed at which the wave group moves, called the *group velocity*, is half the speed of the dominant individual waves. Ongoing, highly mathematical studies of wave groups and the concept of "groupiness" may lead to an improved understanding of the interactions of waves and the shore.

But individual waves or wave groups cannot exist forever. The tendency is for water to seek its own level—for ripples to return to a calm surface. Waves decay or attenuate through either of two general processes: dispersion or dissipation.

A simple way to imagine dispersion is to visualize a series of waves traversing a long, deep channel that suddenly opens into a wide, deep bay. Diffraction causes the waves to spread to the sides as well as to continue advancing in line with the channel. As a result, the wave front tends to become semicircular, and the overall wave heights diminish (figure 3.7). As the waves expand into an increasing area as the semicircle widens, the energy density, and hence the wave height, decrease. Energy density is units of energy per unit area. Waves also disperse as they move away from the origin through an unrestricted body of water. As the front broadens into calmer water, both the energy density and the wave height decrease.

Wave dissipation is more complex. While gravity is a major restorative force that tends to level the sea surface, other factors consume the energy of the waves. Water is not a frictionless fluid. The simple acts of moving water and overcoming inertia consume some of the wave energy. The viscosity of water also contributes to the dissipation of wave energy, as does turbulence. In shallow-water waves, friction between the moving water and the bottom is another source of energy loss.

Inevitably, moving water interacts with the bottom in many ways. Consider the simplest case of a very smooth bottom and a smooth, or laminar, current. The current velocity has to drop from whatever it might be near the surface to very near zero at the seabed. Most of the changes take place near the bottom in a region called the bottom *boundary layer*. The thickness of the boundary layer depends on the velocity of the current, the viscosity of the water, and the roughness of the bottom. The current velocity increases logarithmically away from the boundary; that is, the change in

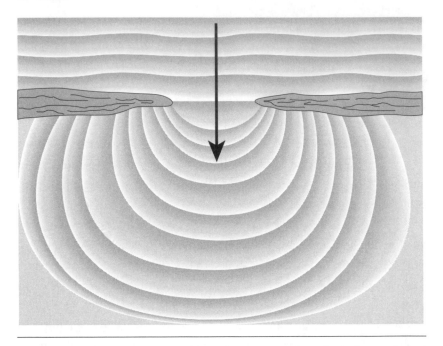

Figure 3.7 Waves disperse as they pass from the open ocean through an inlet or a channel that opens into a wide bay. The wave front expands through diffraction to fill the bay, and the wave height decreases away from the inlet. (Illustration redrawn by Network Graphics from the original by P. L. Mason)

speed is great close to the boundary, and the rate of increase slows away from the boundary. The nature of the bottom boundary layer, along with the overall current velocity and the character of the sediment, is a factor in moving sediments.

* * *

Remember that steep waves break when they become unstable. Classification systems of breaking waves list four types of breakers: spilling, plunging, surging, and collapsing (figure 3.8). The slope of the nearshore zone and the steepness of the wave just before it reaches shallow water determine the type of breaker.

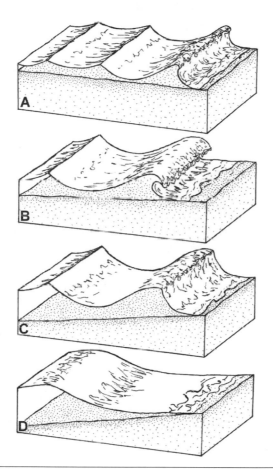

Figure 3.8 The four types of breakers: (A) spilling; (B) plunging; (C) collapsing; (D) surging. (Illustration by P. L. Mason)

A breaker starts to spill when it becomes too steep and the top flows down the front of the wave. Spilling breakers tend to occur where the seafloor under the surf zone has a very gentle slope. Spilling takes place over a relatively long distance. Indeed, the process is so slow that at times the wave hardly appears to be breaking.

Picture waves for surfing. What most people call "breakers" technically are plunging waves. The seafloor under plunging breakers is steeper than that under spilling waves. As the wave shoals, its crest curls forward and breaks in a single crash. After a plunging wave breaks, the water contin-ues to push forward, up the beach. When a wave approaches the breaker zone at an angle, the break progresses along the wave front as that part of the wave reaches the shallower depth. Surfers enjoy this breaker because it is long in both distance and time.

Although a surging wave is characterized as a breaker, its crest does not really break. On a very steep beach, the crest does not actually fall over but peaks, rapidly increases in height, and surges or rushes up the beach face without spilling or plunging.

Recent classifications place the collapsing wave as an intermediate form between plunging and surging breakers. The wave crest itself does not break, but the lower part of the front face steepens and falls. The re-sult is a turbulent, foamy surface that moves up the beach.

The story may not end when a wave breaks. If it breaks on an offshore bar inshore of which the water deepens, the wave can re-form and con-tinue toward the coast, where it might break again. This is the process that generates the "outer" and "inner" breaks. The re-formed "inner" waves seldom develop as cleanly as their offshore "outer" parents. Long-shore currents can be quite strong in the water between the bar and the shore. The interactions among the re-formed waves, longshore currents, backwash from the waves that reach the shore, and other associated pro-cesses produce a complex and dynamic environment.

Watching waves break can be both hypnotic and exciting. The contin-ual repetition, especially if the waves are slow (long period), can lull us to sleep. The obvious energy and drama of huge, breaking waves invigorates almost everyone who sees them. Watching waves break also can help us understand them. One time while vacationing at a beach, I was sitting on a low balcony that had a great view. I put down the novel I had been reading and just watched the low swell as it broke on the beach. It was near high tide, the wind was calm, and the waves were very clean—coming from only one direction and at a constant period. After a short time, something drew my attention to the waves as they were reflected from the beach face. Because the incident waves approached the beach at a slight angle, they reflected outward at a similar, small angle. The conditions were per-fect for seeing a reflected wave refract (bend) as it moved into the slightly deeper water. As the wave moved into deeper water, it sped up so that its crest was almost perpendicular to the shoreline—exactly the reverse of the way we usually think of refraction causing incoming wave crests to bend so that they almost parallel the beach. As I watched, a wave moving along the shore occasionally interacted with an incoming breaker so that a definite "high spot" ran along the crest of the breaker where it combined with the crest of the (now) shore-parallel reflected wave. It was a perfect natural laboratory for observing and learning about the complex interactions of waves with one another and with the seafloor.

* * *

Tsunamis (Japanese for "harbor wave"), often incorrectly called tidal waves, are long-period waves (hence the use of the term "tidal") generated by a submarine earthquake, volcanic eruption, or similar event. The confusion in terms stems from the fact both tides and tsunamis are (very) long-period waves and, as such, have some characteristics in common. Some wave classifications based on wave period lump the longer-period waves together as "tidal waves." The tsunami of December 26, 2004, which devastated the coasts of Indonesia and other nations around the Indian Ocean, raised the worldwide awareness of this potentially catastrophic phenomenon.

Like tides, tsunamis traverse the seas as shallow-water waves. Remember that a wave in water whose depth is less than 5 percent of the wave length is a shallow-water wave; tsunamis have very long wave lengths. In the open ocean, they have very little height, but as they move onto the shallows of the nearshore, the same processes that cause wind waves to build and break operate on tsunamis. As a result, a tsunami can be many feet high when it reaches the shore. Tsunamis cross oceans quickly. Remember the equation for the speed of a shallow-water wave:

$$C = \sqrt{(gh)}$$

where g is gravity and h, water depth. The average depth of the Indian Ocean is about 12,700 feet (3,870 m), so the speed of a shallow-water wave is about 435 miles per hour (700 km/hr). Even in only 100 feet (30 m) of water, a shallow-water wave moves at about 38 mph (60 km/hr). These high speeds make it important to have an efficient tsunami detection-and-warning system.

The tsunami caused by the catastrophic earthquake just offshore of northeastern Japan on March 11, 2011, was a superb teaching tool. The enhancements to the Pacific Tsunami Warning System that were made following the 2004 tsunami in the Indian Ocean worked, so there was plenty of time to prepare as the tsunami charged east across the Pacific toward Hawaii and the Americas. However, even though tsunami warnings were broadcast for the northeastern coast of Japan, the epicenter of the earthquake was so close to shore that the warning time was very short—too short for many people to find refuge

The tsunami was forecast to reach the Hawaiian Islands at a little after 3:00 A.M., Hawaii time (8:00 A.M., Eastern Standard Time). Television stations in Hawaii streamed live broadcasts on the Web. I learned a lot from watching two sites. One was an open-ocean beach. Even with the poor lighting, it was easy to see fairly small waves hitting the beach at what looked like low tide. Then over a period of perhaps 5 minutes, the water level got higher and higher until it was at the top of the berm, still with small waves. A short time later, the water level began to fall until it was lower than it had been at the start of the sequence. The broadcaster commented that he had surfed at that beach many times and had cut his feet on the shells a few times, but had never before seen the bottom exposed so far out. There was not any palpable drama to this tsunami in this place, no great breaking wave. But what happened was important: the water of a full tidal cycle—really more, since the water level appeared to reach above normal tide levels—moved during half an hour instead of half a day.

This explains the action at the second site, a marina in a protected area. The currents flowing through the marina were so great that they pulled boats from their moorings and ripped apart some docks. Something near the volume of water that would cycle through the marina during a low–high–low tidal cycle moved in during half an hour instead of half a day, so the currents had to be extreme.

Usually, there was no indication of the tsunami as a breaking wave. The video of the tsunami inundating the airport at Sendai, Japan, shows a very rapidly advancing, ever-thickening sheet of water enveloping the ground.

BOX 3.4

One cubic foot (0.03 m³) of seawater, about 8 gallons (38 liters), weighs approximately 64 pounds (29 kg). So 1 cubic foot of seawater in a wave in water with a depth of 5 feet (1.5 m) (a shallow-water wave)—using the equation $C = \sqrt{(gh)}$—has a speed of about 12 feet per second (8 miles per hour [13 km/hr]) and exerts about 768 pound feet per second, or about 1.4 horsepower (1 horsepower is 550 pound feet per second).* And there are many, many cubic feet of water in every wave.

---

*Horsepower is a traditional British unit of measure, and the metric system, more properly the Système International (SI), uses watts. As SI is increasingly used around the world, the conversion is 1 horsepower equals about 750 watts, or 0.75 kilowatt, and 1 kilowatt equals 1.34 horsepower.

Contrary to what the movies would have us believe, tsunamis, with their huge wave lengths, are unlikely to reach an elevation one-seventh of their lengths. In some circumstances, the interactions of the tsunami, the currents, and the bottom do result in a breaking wave, but the effects of the immense surge of water in a tsunami can be devastating, even if it does not break.

As the tragic day of the earthquake progressed, it was interesting to scan the graphs of the observed water levels at various stations from Hawaii, the Aleutian Islands, and the West Coast of North America to see the tsunami travel from west to east. The graphs also depicted the many subsequent, usually smaller, tsunamis that followed the initial wave.

* * *

This chapter began with the statements "Waves move energy from one place to another" and "Waves that break on the beach and rock boats also erode the shore and modify beaches." In other words, waves do the work of erosion. Physicists define work as both the energy expended by natural phenomena and the force applied to an object with resulting motion.

BOX 3.5

Here, in the discussion of waves, is a good place to mention physical properties of seawater. One cubic foot (0.03 m³) of seawater weighs close to 64 pounds (29 kg). The density of seawater is about 0.6 ounce per cubic inch (1.02–1.03 g/cm³), which is equivalent to 1,719 to 1,736 pounds per cubic yard (1,020–1,030 kg/m³). The density varies slightly with temperature, salinity, and pressure (depth). The average salinity of the ocean is 35 psu (Practical Salinity Units), which, essentially, is the same as 35 parts per thousand (ppt) dissolved salts. Another way to look at 35 psu is that there are 35 grams (1.2 oz) of salt mixed into 965 grams (34 oz) of water, for a total of 1,000 grams (35 oz). Average seawater is about 3.5 percent salt by weight. In an estuary, where freshwater dilutes seawater, the salinity is less. Indeed, because the differences in the densities of freshwater and seawater come from the differences in salinity, some estuaries exhibit distinct layering, or *stratification*.

In tropical lagoons and similar environments with little or no inflow of fresh water and lots of evaporation, the salinity can increase substantially. In Laguna Madre, on the Texas coast of the Gulf of Mexico behind Padre Island, salinity can reach 45 psu. Little freshwater flows into the lagoon, which has an extremely restricted connection to the open Gulf, so during hot and dry periods, much of the water is lost to evaporation, but the salts remain.

Power is the quantity of work accomplished in a unit of time. As discussed in chapter 2, force is mass times the square of the velocity:

$$F = mv^2$$

where F is the force; m, the mass; and v, the velocity (the terms "velocity" and "celerity" are interchangeable). Seawater has substantial mass, so waves slamming against a structure carry a lot of force and can do major damage. Understanding the force that waves can exert on the beach is essential to understanding how beaches change.

4

TIDES

In studying tides, we have to think of the sun and the moon. Tides are a product of the gravitational attractions between Earth and the moon and between Earth and the sun. We also have to switch our thinking about the arrangement of the solar system back to that of our ancient ancestors and imagine Earth as being the center of the universe. The rotation of Earth on its axis and the revolutions of the moon and the sun "around" Earth cause the tides. The three-dimensional shapes of the ocean basins further modify the tides.

First, we take Isaac Newton's Law of Gravitation:

$$F = G \, (M_1 M_2 / r^2)$$

where F is the attractive force between two bodies; G, the universal gravitational constant; M_1 and M_2, the masses of the two bodies; and r, the distance between them. By plugging in the numbers, we can calculate that the attractive force between Earth and the moon is almost twice as strong as that between Earth and the sun. The centrifugal forces resulting from the revolution of these celestial bodies around one another balance these attractive forces.

Earth and the moon rotate on their own individual axes, and they revolve around the center of gravity of the combined Earth–moon system. Because Earth is so much larger, more massive, than the moon, the com-

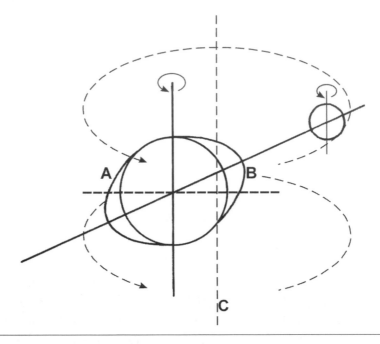

Figure 4.1 The Earth–moon system influences tides. The moon and Earth each rotate on their individual axes while the system also rotates on a separate axis (C) through its center of mass. The water surface on Earth bulges in response to the gravitational forcing toward the moon and the centrifugal force on the outside of the rotation. As Earth rotates on its own beneath the bulges, a spot on its surface passes from beneath a thin part of the bulge (A), lower high water; out of the bulge; then back under a thick portion of the opposite bulge (B), higher high water; out of the bulge; and back to its original position. (The drawing is not to scale, and the angle of the moon's orbit to Earth's axis is exaggerated.) (Illustration by P. L. Mason)

mon center of gravity of the two-body system is within the Earth, about 1,000 miles (1,700 km) beneath the surface (figure 4.1). Water is a fluid and, thus, is free to change shape in response to the forces applied to it. If the moon's gravitational force had the greatest influence on Earth's surface water closest to the moon, theoretically the water would bulge out. The centrifugal force would be strongest at the spot directly on the exact opposite side of the globe. Water particles not in these two places would experience slightly different forces that generally decrease with distance from the points of maximum force. In the simplest case, if all else were equal, a single, uniformly deep ocean would completely cover Earth.

Unfortunately, the tide is not that simple. The moon's gravity does affect everything on Earth, but its influence is so small compared with Earth's own gravity that it is insignificant. The National Ocean Survey estimates that, on average, the attraction of the moon's gravity "is only about one 9-millionth part of the force of earth-gravity" and is entirely insufficient to elevate the water's surface. What does cause the tides is the horizontal component of the moon's gravity, tangential to Earth's surface, which pulls the water toward the point of maximum traction and toward the opposing point of maximum centrifugal force. According to the National Ocean Survey's *Our Restless Tides*:

> At any point on the earth's surface, the tidal force produced by the moon's gravitational attraction may be separated or "resolved" into two components of force—one in the vertical, or perpendicular to the earth's surface—the other horizontal or tangent to the earth's surface. This second component, known as the tractive ("drawing") component of force, is the actual mechanism for producing the tides. The force is zero at the points on the earth's surface directly beneath and on the opposite side of the earth from the moon (since in these positions, the lunar gravitational force is exerted in the vertical—i.e., opposed to, and in the direction of the earth-gravity, respectively). Any water accumulated in these locations by tractive flow from other points on the earth's surface tends to remain in a stable configuration, or tidal "bulge."
>
> Thus there exists an active tendency for water to be drawn from other points on the earth's surface toward the sublunar point and its antipodal point and to be heaped at these points in two tidal bulges. Within a band around the earth at all points 90° from the sublunar point, the horizontal or tractive force of the moon's gravitation is also zero, since the entire tide-producing force is directed vertically inward.

The rotation of Earth on its axis should bring each point on Earth's surface under a portion of each of the two bulges once a day, 12 hours apart, in the two daily high tides. These are called *semi-diurnal tides*. But the axis of rotation tilts away from perpendicular to the Earth–moon connecting line. Thus a point on Earth's surface will not pass underneath corresponding thicknesses of the two bulges (see figure 4.1), with the consequence that successive high tides have different elevations—a higher high tide followed by a lower high tide followed by a higher high tide. The difference between the two elevations is the *diurnal inequality*. The same thing happens with low tides: they occur halfway in distance between the bulges that represent the high tides and halfway in time, 6 hours, between the times of high tide.

But anyone who has spent time at the shore has noted that the time of high and low tides changes each day. It takes 24 hours, 50 minutes for the moon to be directly "over" the same point on Earth. This is because the moon orbits Earth at an angular rate that is slightly different from that of Earth's rotation on its own axis. So the ideal, totally ocean-covered Earth has to rotate slightly more than a full circle, slightly more than 360 degrees, to catch up to the bulge. Hence the tides occur about 50 minutes later each day.

There are further lunar complications that may seem lunatic. The period of the moon's orbit around Earth is approximately 27.3 days, which is called the *sidereal month*. But as Earth orbits the sun, it takes approximately 29.5 days—a *synodic month*—for the moon to complete a full revolution with respect to its orientation with the sun. The moon's elliptical orbit brings it closer to and farther from Earth; for example, in 2010, the closest *perigee* was 221,576 miles (356,592 km), and the farthest *apogee* was 252,613 miles (406,541 km). Also, the plane of the moon's orbit around Earth is tilted about 5 degrees from the plane of Earth's orbit around the sun.

Now let us consider the sun's influence on tides. The tide-producing force of the sun is about 40 percent that of the moon, so it is not insignificant. The geometry of the relationships between Earth and the sun is similar to that between Earth and the moon. Included among these relationships are Earth's yearly revolution around the sun, the tilt of Earth's rotational axis relative to the plane of the orbit, and the variation between perigee and apogee of the orbit. The solar-generated tides are added to (or subtracted from) the lunar-generated tides.

The combination of the lunar and solar relationships with Earth are visible as the "monthly" cycle of new moon, waxing moon, full moon, and waning moon and the accompanying two-week variation in *tidal*

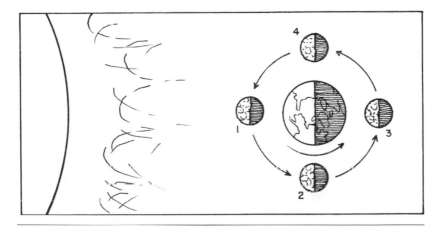

Figure 4.2 The phase of the moon affects tides: (1) the sun, the moon, and Earth are aligned, a configuration known as syzygy, and the tidal range is enhanced in spring tides; (2) the sun, the moon, and Earth are in quadrature; the phase of the moon is the first quarter and waxing, and the tidal range is diminished to the neap range; (3) the arrangement again is syzygy with spring tides, but the moon is full; (4) the configuration again is quadrature with neap tides, but the moon is in the last quarter and waning. In position 1, if the alignment is perfect, the moon completely blocks the sun in a solar eclipse; in position 3, if the alignment is perfect, Earth's shadow covers the moon in a lunar eclipse. (The drawing is not to scale.) (Illustration by P. L. Mason)

range, the difference in elevation between successive high and low tides (figure 4.2). When the moon is directly between Earth and the sun, a situation known as *syzygy*,* the side of the moon facing Earth is in shadow. We see a new moon; the separate gravitational forces of the sun and moon are aligned; and the solar and lunar tides add together, yielding a higher than average tide, called a *spring tide*. (If the sun–moon–Earth alignment is "perfect," the moon will cover the disk of the sun and we will marvel at a solar eclipse.) About a week later, when the moon has moved 90 de-

* "Syzygy" is a word known to crossword-puzzle enthusiasts. My most interesting connection with the word occurred many years ago when I was being "qualified" as an expert witness in a murder trial to provide information about the likely water level at the time and place a woman died. One of the trial attorneys, I think the prosecutor, asked me to spell and define "syzygy." When I did so quickly and to his satisfaction, I was accepted as an "expert" witness.

grees through its orbit around Earth, one-half of the side of the moon facing Earth is sunlit, and one-half is in its own shadow. This arrangement is termed *quadrature*, and we see a first-quarter moon. Because the lunar and solar tide-producing forces are at right angles to each other, the tides are out of phase with each other, yielding a diminutive tide called a *neap tide*. Another week later, the sun, Earth, and moon line up again for a syzygy. But Earth is directly between the sun and the moon, so the side of the moon facing Earth is completely sunlit. We see a full moon, and again there is a spring tide. (If the sun–Earth–moon alignment is "perfect," the moon will be in the shadow of Earth, and we will enjoy a lunar eclipse.) A week later, the moon has traveled another 90 degrees through its orbit around Earth. The moon again is in quadrature, the other half of the side of the moon facing Earth is sunlit, we see a last-quarter moon, and there is a neap tide.

The relatively large variation in tide-producing forces associated with the fortnightly, spring–neap cycle is superimposed on the much lesser lunar orbital (perigee–apogee) cycle, so spring tides sometimes coincide with lunar perigean tides. Every few years, the media hype extraordinarily high tides when spring and perigean tides converge. Although the consequences of the astronomical alignment are real, they are far from catastrophic and are not worthy of the inordinate attention paid in the media. NOAA's "What Are the 'Perigean Spring Tides'?" describes the combination of spring tides and lunar perigean tides:

> The difference between the "perigean spring tides" and the normal tidal ranges for all areas of the coast is small. In most cases the difference is only a couple of inches. The worst case that we have found occurs in certain areas of the Alaska coast where the range of the tide was increased by approximately 6 inches. But when you consider that these areas have a average tidal range of more than 30 feet, the increase is but a small percentage of the whole (less than a 2% increase).

The effect of solar perigean tides also is quite small. Indeed, if you look at the tables of predicted tides, "perigean spring tides" are difficult to distinguish from the more frequent cycles of conventional spring tides.

Even on the perfect, ocean-covered Earth, predicting tides would be difficult because the astronomical relationships among the sun, moon, and Earth cycle with different periodicities. Fortunately, mathematicians can compute the complex, harmonic analyses needed to determine the relative importance of each and the sum of the many factors at virtually any time. More than 30 constituents are used to predict tides. As it turns out, the combined cycle of the major components has a period of 18.61 years, known as a *tidal epoch*. However, the water level observed at any particular spot at any particular time is the result of the astronomical tide-producing forces, barometric pressure, wind, and water temperature. Thus in order to assess changes in sea level through time, reference elevations such as mean low water, *mean tide level* (MTL), and mean higher high water are calculated on a 19-year interval. The extension from 18.61 to 19 years both simplifies computation and, by encompassing equal numbers of the full year's seasons, more properly includes seasonal factors. Thus when specifying an elevation measured from a tidal datum, it is necessary to specify the 19-year epoch for which the datum was determined because as sea level changes through time, the elevation reached by tides changes, so the tidal datums also change.

* * *

The discussion of tides becomes more complex when it shifts from the uniformly water-covered, continent-free globe to the real world. The continents break the ocean into several basins, and the water depth varies within and between them. Think of the potential wave length of a tide. On the all-ocean Earth, the entire globe would constitute only two tidal cycles. The wave length of the tidal wave (not to be confused with the wave length of a tsunami) would approach one-half of Earth's circumference. Since most ocean basins experience two tidal cycles in a day, it is not a great leap to assume that the wave length would approach the size of the ocean basin. And as an ocean basin likely is more than 20 times wider than it is deep, tidal waves always are shallow-water waves. In the Pacific Ocean near Japan, the Challenger Deep (the deepest known point in the oceans) is 6.86 miles (11 km) deep. Twenty times that depth is 137 miles

(220 km). Remember that a shallow-water wave is one for which the water depth is less than 5 percent of the wave length. And the speed of a shallow-water wave is

$$C = \sqrt{(gh)}$$

where C is the wave speed; h, the water depth, and g, the acceleration of gravity. Therefore, as a tidal wave moves across a shallowing continental shelf or up a river or an estuary, by and large its speed progressively slows. But the wave period (and its inverse, the wave frequency) remains unchanged. So depending on the depth and length of a tidal river, only part of a tidal wave, a complete wave, or multiple waves may be moving up the estuary.

The rotation of Earth on its axis also affects the tides. The Coriolis parameter, often and incorrectly called the Coriolis force, deflects motion toward the right of its path in the Northern Hemisphere and toward the left in the Southern Hemisphere. This diversion adds a rotary motion to the tides as a tidal wave traverses an ocean. Rotary tidal systems are called *amphidromic systems* and cycle counter-clockwise in the Northern Hemisphere. In an amphidromic system, high tide is like a spoke on a bicycle wheel. The tidal range approaches zero at the wheel's hub, the center of the rotation, called the *amphidromic point*, and increases away from the center.

To visualize an amphidromic system, think of a huge pan with a very shallow layer of water. If you lift one side of the pan and put it back down, the water sloshes back and forth. But if the pan is big enough for the Coriolis parameter to affect the flow, instead of simply moving from one side of the pan to the other, the water moves slightly toward the right side as it sloshes toward the far end and slightly toward the left side as it flows back to the start. High and low water circle the sides of the pan, while the water level in the center changes very little. This circular flow is similar to an amphidromic system.

Because the relative importance of many of the tide-producing forces depends on specific spatial relationships—for example, the point "directly under the moon"—the "shape" of the tidal wave varies from place to place on Earth.

* * *

Before proceeding farther, it is necessary define a handful of new terms and remember that there is a difference between the change in elevation of the water surface that we call the tide and the (tidal) current that both affects and is a consequence of the tide.

High water (high tide), low water (low tide), and mean (or average) tide level are self-defined and refer to the water level/tide level. The tidal range is the difference in height between successive highs and lows. In the more common situation where there are two high and two low tides each day, the tidal range usually cycles greater–lesser–greater–lesser. The difference between a day's tidal ranges is called the diurnal inequality. And because the two high tides in a day usually differ in elevation, as do the two low tides (how else could there be a diurnal inequality?), there are terms to describe the daily extremes: "higher high water" and "lower low water." The corresponding terms for the more moderate of a day's tides, "lower high water" and "higher low water," seldom are used because their importance to society is relatively minor.

Vessel operators need to be certain that the water is deep enough for the draft of their ships, so the predicted elevation of lower low water relative to a standard, reference elevation (datum) may be critical for safe passage. Conversely, a captain who plans to sail his vessel under a bridge would be most concerned with the worst case of higher high water. Similarly, people who are anxious about coastal flooding more likely would be concerned about the predicted daily extreme of higher high water than with a lesser elevation.

The most common type of tide is the semi-diurnal tide. The tidal "day" is about 24 hours, 50 minutes, so sometimes there are only three high or low tides—rather than the semi-diurnal two highs and two lows—in a calendar day. For example, if high tide occurs just before midnight, low tide would be about 6:05 A.M., the next high tide close to 12:20 P.M., the next low tide around 6:30 P.M., and the fourth tide of the cycle at 12:40 A.M. the following day.

If the diurnal inequality is relatively great, the tide is called a *mixed tide* (figure 4.3). In some locations, the conditions are such that there is only one tidal cycle each day: one high and one low tide This is called a *diurnal tide.*

Worldwide, the tidal ranges vary considerably. Lakes have no discernable tides, and the effects of wind and barometric pressure overwhelm whatever the astronomical tide might be. The tidal ranges in the Mediterranean Sea and the Gulf of Mexico are very low, about 1 foot (30 cm), whereas in the Bay of Fundy, between the Canadian provinces of New Brunswick and Nova Scotia, the tidal range can exceed 50 feet (15 m). Tides ranges of 0 to 6.5 feet (0–2 m) are called microtidal; 6.5 to 13 feet (2–4 m), mesotidal; and greater than 13 feet (4 m), macrotidal.

In addition to the theoretical relationships of the tide-generating forces, other factors work to modify the tidal range in specific areas. Visualize an opening in the coastline that separates the ocean from an embayment. During half a tidal cycle, the time it takes the tide to rise or to fall, a definable volume of water passes through the inlet. This volume is called the *tidal prism* and is a function of the tidal range, the rate of change of the tidal elevation, the current velocity (as it changes through time), and the cross-sectional area of the inlet. Ignoring the diurnal inequality and the freshwater flowing into the embayment, the flood and ebb prisms should be identical. Alternately, the tidal prism can be calculated as the quantity of water required to fill the space in the basin between the levels of low and high tide.

A wide variety of tidal-prism and basin relationships affect tidal flow. If the inlet is relatively wide and the basin is relatively long and narrow, like a river, the tide-producing forces push the tide upstream. The hydraulic head (water at a higher level flowing toward lower water) also propels the tide upstream. This situation is fine as long as the river has a uniform cross section nearly the same as that of the inlet. If the riverbed rises in elevation upstream, however, the fixed volume of water in the prism must rise to compensate for the change in the trough, or channel, volume. Similarly, if the basin narrows, the constant volume of water has to rise

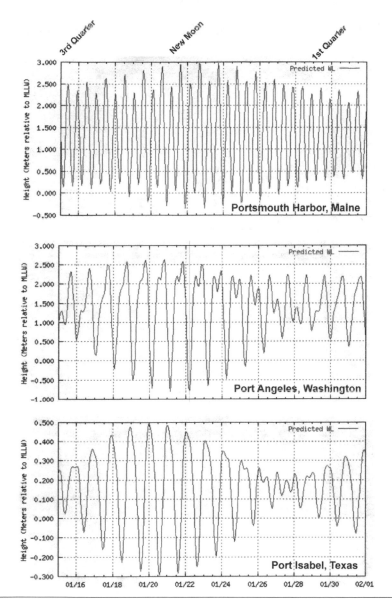

Figure 4.3 Spring–neap tide cycles in late January 2004 at three different locations: Portsmouth Harbor, Maine, displays semi-diurnal tides with a relatively small diurnal inequality; Port Angeles, Washington, has mixed tides with a pronounced diurnal inequality such that there is only one significant tidal cycle each day; Port Isabel, Texas, has diurnal tides except during a brief period of mixed tides around neap tides. (Note that the Port Isabel graph has a different vertical scale from the other graphs.) (From National Oceanographic and Atmospheric Administration, Center for Operational Oceanographic Products and Services, http://co-ops.nos.noaa.gov)

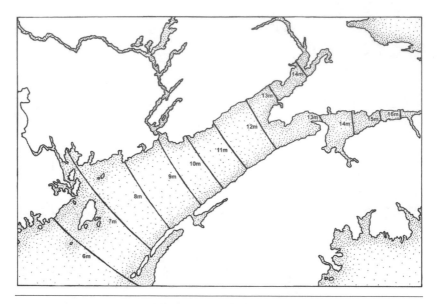

Figure 4.4 The lines of equal range (co-range lines) for spring tides in the Bay of Fundy. (Modified from Bay of Fundy, http://www.bayoffundy.com/mariner/images/tidalmapbig .jpg)

because it is being squeezed from the sides. So in a long, narrow, wedge-shaped basin that is open to the ocean at its widest and deepest aspect, the tidal range will increase up-basin as the shoaling floor and narrowing sides constrict the flood tidal prism. The Bay of Fundy has such an up-basin increase in tidal range (figure 4.4).

The mouth of the Bay of Fundy is roughly 80 miles (130 km) across, and the bay narrows to about half that width before it splits into two arms; Minas Basin–Cobequid Bay and Chignecto Bay. It is not surprising, then, that the tidal range roughly doubles as the incoming volume of water (tidal prism) squeezes into the narrower space. Figure 4.4 shows spring-tide conditions in the Bay of Fundy, and figure 4.5 displays the average conditions. The spring tides increase in range from 20 to 40 feet (6–12 m), while the average change is from 16 to 33 feet (5–10 m). Charted lines of equal tidal range are called *co-range lines*. Figure 4.5 also shows the progress of the time of high tide up the Bay of Fundy. The lines connecting points where high tide occurs (really, lines connecting points where the

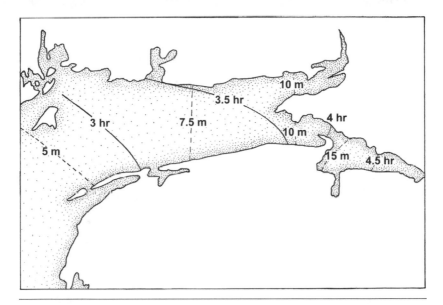

Figure 4.5 The lines of equal tidal range (co-range [solid lines]) and stage (co-tidal [dashed lines]) in the Bay of Fundy. (Modified from P. R. Pinet, *Invitation to Oceanography*, 3rd ed. [Sudbury, Mass.: Jones and Bartlett, 2003], 274)

same stage of the tide occurs) at the same time are termed *co-tidal lines*. In figure 4.5, it appears that the tidal wave moves through the length of the Bay of Fundy in 2.5 hours (box 4.1).

In Fundy, the relationship of wide inlet and narrowing bay amplifies the tidal range. A common situation is a narrow inlet connecting the ocean with a back-barrier bay or lagoon (figure 4.6). In this situation, the tidal inlet serves as a damper that reduces the tidal range in the lagoon as compared with that along the ocean coast. This condition makes sense. During a rising tide, the water has a limited amount of time in which to flow into the lagoon, and the inlet is a pinch valve that limits the rate at which the water can flow from the ocean into the lagoon. When the tidal prism spreads across the whole lagoon, the limited volume is not sufficient to raise the water to the level of the high tide on the immediately adjacent ocean coast. The ocean tide begins to fall before the water level in the lagoon rises to the full elevation of the ocean-side high tide. Lacking the push of a flood current and the differential elevation, or head, the

BOX 4.1

An Aside About "Common Knowledge"

In the Bay of Fundy, according to some sources, a near equality of the tidal period and the resonance period serves to amplify the tidal range up-basin. The resonance period is the length of time it would take a *seiche* to oscillate from one end of a closed basin to the other and back. I have not been able to verify this assertion. Indeed, simple calculations tend to dispute the common lore.

Using figures 4.5 and 4.6 and inspecting tide tables, it appears that the tidal wave takes about 2.5 hours to travel the length of the bay. Rough measurement shows that the tidal wave travels approximately 170 miles (270 km) in that time. Plugging into the equation for the speed of a shallow-water wave,

$$C = (gh)$$

where C is the speed of the wave; g, the acceleration of gravity (32 ft/s^2 [9.8 m/s^2]); and h, the water depth, and solving for water depth,

$$C^2 = gh, \text{ so } h = C^2 / g$$

the wave speed is 170 miles (270 km) in 2.5 hours, or 68 miles per hour (108 km/hr or 108,000 m/hr), which divided by 60 twice (and converting from miles to feet) yields 100 feet per second (30 m/s) as the wave speed. So, if $h = C^2 / g$, the water depth is 310 feet (95 m).

The accompanying NOAA nautical chart, which includes the lower portions of the Bay of Fundy, gives depths in fathoms (1 fathom equals 6 feet [1.8 m]). So 310 feet (95 m) is very close to 50 fathoms. An "eyeball" analysis of the chart indicates that the 50-fathom *isobath* extends one-quarter to one-third of the distance up the bay, so an average depth of 50 fathoms is not unreasonable, but likely is slightly high.

Moving to the equation for the resonance period,

$$T = 2L / \sqrt{(gh)}$$

where T is resonance period (time) and L, length, then

$$T = (2) (170 \times 5,280) / \sqrt{(32 \times 310)}$$

So the resonance period is 18,024 seconds, or 5 hours. In no way does this approach the reputed 11- to 13-hour resonance period.

Using the formula for the oscillation period, C. A. M. King stated that the Bay of Fundy "has a natural period of oscillation of 11.6 to

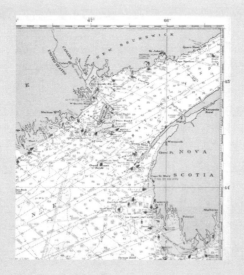

The lower portions of the Bay of Fundy adjacent to the sea. (From National Oceanographic and Atmospheric Administration, Office of Coast Survey, chart 13003 [2010], http://www.charts.noaa.gov/OnLineViewer/13003.shtml)

13 hours, which corresponds closely to the semi-diurnal period."* By working the problem another way, if we accept the resonance period as 12.3 hours (remembering to convert to seconds) and the depth as 310 feet (95 m), and we solve for length,

$$T - 2L / \sqrt{(gh)}, \text{ thus } L = T\sqrt{(gh)} / 2$$

which equals 418 miles (673 km). The Bay of Fundy clearly is not that long. Or solving for depth and using 12.3 hours and a length of 170 miles (270 km),

$$h = 4L^2 / T^2g, \text{ thus } h = 51 \text{ feet (16 m)}$$

which appears to be far too shallow a depth.

According to the West Nova Eco Site, "The Bay of Fundy is 400 km [250 miles] long with an average depth of 75 m [245 ft]."[†] Using these numbers, the wave speed

$$C = \sqrt{(gh)}$$

calculates as

$$C = (32 \text{ ft/s}^2 \times 245 \text{ ft})^{1/2} [(9.8 \text{ m/s}^2 \times 75 \text{ m})^{1/2} \text{ or } 27 \text{ m/s}]$$

BOX 4.1

(*continued*)

which is 60 miles per hour (97 km/hr.) This is pretty close to the 68 miles per hour (108 km/hr) calculated earlier. But solving for the period, $T = 2L / \sqrt{(gh)}$,

$(2)(250 \text{ miles}) / ((32)(245))^{\frac{1}{2}} = (2,640,000 \text{ ft}) / (88 \text{ ft/s}) = 30,000$ seconds, or 8.3 hours $[(2)(400 \text{ km}) / (27 \text{ m/s}) = (800,000 \text{ m}) / (27 \text{ m/s}) = 29,630$ seconds, or 8.2 hours]

about 50 percent different from the given resonance period of about 12.3 hours.

The West Nova Eco site also states, "The tide moves 105 cubic km or 3.5 million cubic feet of water in and out of the Bay of Fundy,"[‡] which clearly equates the two volumes. Unfortunately, they are not equal. Doing the calculations, 1 cubic kilometer is 1,000 meters × 1,000 meters × 1,000 meters, and 1 meter equals 3.28 feet, so 3,280 feet × 3,280 feet × 3,280 feet yields 35,288 million cubic feet, even before counting the factor of 105, which would raise the number to 3.7×10^{12} cubic feet—a substantial error in calculation. So do not believe everything in print or on the Web, or, as was stated in a much different context, "Trust but verify."

If we do compare the change in width of the Bay of Fundy to the change in tidal range, there appears to be a simple inverse relationship. Using figures 4.5 and 4.6, the width of the bay at the 33-foot (10-m) co-range line is very nearly one-half the width at the 16-foot (5-m) co-range line. This general observation suggests that the geographic constriction of the narrowing, and perhaps of the shoaling, bay has a much greater effect than similar "resonance frequencies."

---

[*]C. A. M. King, *Beaches and Coasts* (New York: St. Martin's Press, 1972), 140–143.

[†]West Nova Eco Site, Climate and Tides, http://www.collectionscanada.gc.ca/eppp-archive/100/200/301/ic/can_digital_collections/west_nova/climate.html.

[‡]Ibid.

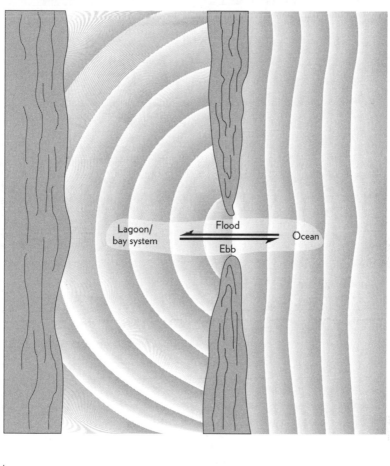

Lagoon/bay system   Flood   Ocean   Ebb

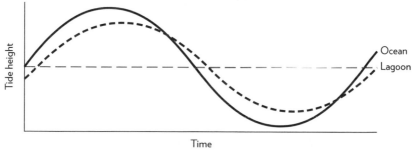

Tide height

Ocean
Lagoon

Time

Figure 4.6 A general representation of the damping of the tidal range by an inlet. The tidal range in the lagoon/bay is lower than in the ocean, and the times of high and low tide are slightly later. (Illustration redrawn by Network Graphics from the original by P. L. Mason)

water level in the lagoon stops rising and, when the level outside the inlet is lower than that inside, begins to fall. The same thing happens with the falling tide: the lagoon cannot empty sufficiently fast, and the water level never drops to that of the low tide on the ocean coast.

Marine scientists and coastal engineers explore the complex relationships of tidal inlets and their associated bays or lagoons. The interactions among the cross-sectional area of a tidal inlet, the (volume of the) tidal prism, and the size of the lagoon are well defined. The size of the cross section of the inlet's throat is tied to the volume of water that passes through it. The relationship is strong enough that the inlet changes size in response to the difference between spring and neap tides.

The relationship between the tidal prism and the cross-sectional area of the inlet, which can be expressed as a mathematical formula, addresses the potential stability of both new and modified inlets. A simple example would be dredging an existing inlet so deeper-draft ships could pass through it. To maintain a constant cross section, the dredged inlet would have to be made narrower to accommodate the greater depth. Also, if bridge pilings, for example, obstruct the inlet, it will deepen to compensate for the blockage. Engineers must specify that the pilings be driven sufficiently far into the bottom to accept both the consequent scour and deepening, or the bridge could become unstable.

A storm may naturally create a new inlet or modify an existing one. In his excellent book *A Celebration of the World's Barrier Islands*, Orrin Pilkey noted a storm-generated tidal prism that widened Oregon Inlet, on the coast of North Carolina, from 0.5 mile (0.8 km) to 2 miles (3.2 km) and deepened it from 15 to 60 feet (4.5–18 m) in a few hours. If a storm opens a new inlet, its longevity and stability depend on how much tidal prism it captures from existing inlets. And if the new inlet steals enough water volume to become stable, tidal prisms and cross-sectional areas of nearby inlets will drop and the inlets may become unstable and close. Even if they stay open, their new depths may be too shallow to allow the passage of vessels that previously traversed them, thereby harming local economies.

Tidal range has another effect on barrier island–inlet systems. In regions with a low tidal range, barrier islands tend to be long and inlets widely spaced; in regions with a high tidal range, the reverse is true. The tidal range along the Texas coast of the Gulf of Mexico is very low, usually less than 1.5 feet (45 cm), and the distance between inlets is great. Padre Island, Texas, is the world's longest barrier island, about 130 miles (210 km) long. Along the southeastern coast of the United States, the tidal range increases from around 3.2 feet (1.1 m) near Kitty Hawk, North Carolina, to just over 7 feet (2.1 m) on the South Carolina–Georgia border. Along the same stretch of coast, the barrier islands change from the long islands of the Outer Banks of North Carolina to the much shorter islands of lower South Carolina and Georgia.

* * *

Tides are an important part of the coastal experience, and understanding what causes them can enhance that experience. Knowing that the tidal range is greater when the moon is new or full than when it is in its first or last quarter can help us assess the likely impact of a storm on the shore. Tidal range is a major determinant of the height of the beach face and the distance between tidal inlets. Knowing that an inlet restricts the flow of water helps explain why the tidal range in a lagoon is less than that along the open coast. Similarly, insight about the shape of a tidal river's channel can aid our comprehension about how and why the tidal range changes up-river. Realizing that high tide occurs roughly 50 minutes later from day to day may guide us in deciding what time we go to the beach on successive vacation days.

5
SEDIMENTS

Sediments collectively are the many solid particles of gravel, sand, silt, and clay. According to the *Glossary of Geology*, sediments originate from the weathering of rocks and are deposited by air, water, or ice. They also accumulate by chemical precipitation from solution or from secretion by organisms. They form in layers on Earth's surface at ordinary temperatures in a loose, unconsolidated form. Sediments are the stuff that makes up beaches, sand dunes, and marshes and that forms the seafloor.

As with many collective terms, there are several schemes for classifying sediments. The terms within each scheme can intermix and lose their specificity when used informally. The technical terms for the study of sediments are "sedimentology" and "sedimentary petrology." The properties that form the bases of the various schemes relate to how sediments are formed, how their physical attributes are described, and how they act.

The introductory chapter of Robert Folk's *Petrology of Sedimentary Rocks* is an excellent summary of the classification of sediments. There are two gross classes of sediments: chemical and terrigenous. The chemical class may be further divided into orthochemical and allochemical groups.

Orthochemical sediments form by natural chemical processes in the basin of deposition and exhibit little, if any, consequence of motion. A striking example of orthochemical deposits are the selenite flowers that form in shallow lagoons and *sebkhas*, or salt flats, in arid and hot areas. Selenite is a form of the mineral gypsum, which is a hydrous calcium sulfate ($CaSO_4 \cdot 2H_2O$) that precipitates as seawater evaporates. Another common orthochemical sediment in the coastal zone is bog iron. The minerals goethite ($HFeO_2$) and limonite ($FeO(OH)\, nH_2O$) can precipitate from water and, as the term "bog iron" suggests, commonly occur in bogs or marshes. Limestone, dolomite, and rock salt are orthochemical rocks. Orthochemical deposits often cement or bind other sediments.

Chemical or biochemical processes form allochemical sediments at Earth's surface. Unlike orthochemical sediments, *allochemical* sediments have been moved since they were formed but have remained in the original basin of deposition. Obvious examples are shells and shell fragments (shell hash, or *coquina*). Additionally, tropical and semitropical beaches often have high percentages of calcium carbonate sediments derived from other biological or chemical processes. Particles often found on the seafloor of shallow lagoons in tropical regions are known as *ooliths*.* Sand-size fragments of coral and of red and purple shells of some foraminifera and other organisms commonly form pink beaches in tropical and semitropical areas.[†]

Terrigenous sediments are individual particles or fragments of rocks that were formed outside the basin of deposition and have been transported into the basin as solids. The word "terrigenous" means "earth formed" or "born of the earth." The terms "detrital" and "clastic" are often used synonymously with "terrigenous." The vast bulk of the sediments that compose sand dunes, beaches, and mud flats are terrigenous.

Any particular sedimentary deposit can contain a mixture of the three broad classes of sediments in any proportions. Other classification systems describe the mixtures, but these more detailed classifications have little importance in the general discussion of sediments other than encouraging the observer to contemplate the set of circumstances and processes that resulted in the specific mixture of a specific deposit.

---

* The *o* is variously pronounced *oh*, as in "oh no," or *ooh*, as in "ooh and aah," yielding *oh-oh-lith* or *ooh-ooh-lith*.

[†] The pink color of the beaches sometimes results from the color of some types of algae that can grow on the surface of the sand particles. Also, the "warm" tinge of the rising and setting sun can enhance the pinkishness of the sediments.

As encompassing as this primary classification might be, it does not provide the detail to identify individual sediment samples. What is the difference between the sediments of a sand dune and a mud flat? Some of the differences are clear to the eye. The three most important characteristics used to describe terrigenous sediments are the size of the individual particles, their composition or mineralogy, and their shape.

The size of sediment grains includes two concepts: the outright sizes of the individual grains and the proportional distribution of sizes within a specific deposit. The size range is enormous. On the large end of the spectrum are boulders, which start with a diameter of 256 mm (about 10 in) and go up to many feet. Clay has the smallest particles. The coarsest clays are $\frac{1}{256}$ (0.0039) mm (0.00015 in), and the finest may be as small as 0.00006 mm (0.000002 in).

With this huge range of nine full orders of magnitude, understanding and managing sediment data is complex and cumbersome. There is no one standard method for measuring the full gamut of sizes, so analyzing a grain-size distribution may require a combination of analytical techniques

According to Francis J. Pettijohn's classic textbook *Sedimentary Rocks*, at the turn of the twentieth century, Johan August Udden devised the scale of grain-size classes that is the basis for the terms commonly used today. In the early 1920s, Chester Keeler Wentworth modified Udden's definitions, and this scale remains the standard terminology. Wentworth's classification doubles or halves the grain size, starting with a 1-millimeter particle (table 5.1). This logarithmic scale serves both scientists and engineers, although engineers define the terms a little differently from geologists. In general conversation, the variations are not significant, but anyone reviewing technical reports, such as data for a public hearing about a beach-nourishment project, should be aware of the differences in professional jargon.

* * *

Because the classifications are based on physical dimensions, sediments must be measured in a way appropriate for those dimensions, and the

TABLE 5.1

Sediment Grain-Size Classification and Nomenclature as Commonly Used in the United States

| Geological | | | | | Engineering | |
| Wentworth Classification | | | | | ASTM Unified Soil Classification | |
Class	Subclass	Microns	Millimeters	Millimeters	Class	Millimeters
Boulder				>256	Boulder	>300
Cobble	Large cobble			128–256	Cobble	75–300
Cobble	Small cobble			64–128	Coarse gravel	19–75
Pebble	Very large pebble			32–64	Coarse gravel	
Pebble	Large pebble			16–32	Fine gravel	4.75–19
Pebble	Medium pebble			8–16	Fine gravel	
Pebble	Small pebble			4–8	Fine gravel	
Granule				2–4	Coarse sand	2–4.75
Sand	Very coarse sand			1–2	Medium sand	
Sand	Coarse sand	500–1,000	½–1	0.5–1.0	Medium sand	0.425–2
Sand	Medium sand	250–500	¼–½	0.25–0.5	Fine sand	0.075–0.425

TABLE 5.1

(continued)

| Geological | | | | | Engineering | |
| Wentworth Classification | | | | | ASTM Unified Soil Classification | |
Class	Subclass	Microns	Millimeters	Millimeters	Class	Millimeters
Sand	Fine sand	125–250	$\frac{1}{8}$–$\frac{1}{4}$	0.125–0.25	Fine sand	
Sand	Very fine sand	62.5–125	$\frac{1}{16}$–$\frac{1}{8}$	0.0625–0.125	Fine sand	
Mud	Coarse silt	31–62.5	$\frac{1}{32}$–$\frac{1}{16}$	0.031–0.0625	Fine-grained soil	<0.075
Mud	Medium silt	15.6–31	$\frac{1}{64}$–$\frac{1}{32}$	0.031–0.0156	Fine-grained soil	
Mud	Fine silt	7.8–15.6	$\frac{1}{128}$–$\frac{1}{64}$	0.0078–0.0156	Fine-grained soil	
Mud	Very fine silt	3.9–7.8	$\frac{1}{256}$–$\frac{1}{128}$	0.0039–0.0078	Fine-grained soil	
Mud	Coarse clay	1.95–3.9	$\frac{1}{512}$–$\frac{1}{256}$	0.00195–0.0039	Fine-grained soil	
Mud	Medium clay	0.98–1.95	$\frac{1}{1024}$–$\frac{1}{512}$	0.00098–0.00195	Fine-grained soil	
Mud	Fine clay	0.49–0.98	$\frac{1}{2048}$–$\frac{1}{1024}$	0.00049–0.00098	Fine-grained soil	

Note: Engineering terms are defined by inches or by the U.S. Standard Sieve Series. European and other terms may differ.

tools vary widely. Clearly, you would not measure a boulder with the same tool that you would use for a grain of silt. Very large sediments are measured with calipers. Traditionally, sieves are used for particles ranging from coarse silts through pebbles or small cobbles, although some laboratories now use other technologies. Pipettes or hydrometers are the standard instruments for determining the grain sizes of silts and clays, but, as with the mid-range sediments, new methods are gaining wider acceptance.

For a traditional sieve analysis, after collecting and drying a sediment sample, you physically divide the sample to obtain a representative subsample that is surprisingly small: 1 to 2.5 ounces (30–70 g). You need a bigger subsample if there is a large percentage of coarser particles. You pour the carefully weighed subsample into a stack of sieves with the coarsest sieve on top. At a bare minimum, there should be sieves for 2 millimeters and 1, 0.5, 0.25, 0.125, and 0.625 millimeter, but a more precise size analysis would include three more sieves spaced between each of the minimum set.

Ideally, all the particles would pass through the first sieve and progress through the progressively finer sieves. Then you load the stack of sieves into a machine that simultaneously shakes and taps the set for a standard time, usually 10 or 15 minutes, that is long enough to ensure that at least 99 percent of the material that could pass through each sieve has done so. If you are using a full set of sieves, you may have to break the stack into two or three sections, as the shaker-taper may hold only eight or 10 sieves at a time. The particles that pass through the tightest mesh collect in a pan. Finally, you carefully weigh the contents of each sieve and the pan and tabulate the data, taking care not to damage or distort the sieves because the finer meshes are fragile. The raw data yield the percentages of sediment particles retained in or passed through each sieve. A carefully standardized version of this technique is the core of grain-size analysis for engineering purposes.

Many research laboratories now measure the grain size of sands by determining the particles' fall velocities through a column of water. In these "settling tubes," a small quantity of sediment is released high in a 3- to 6-foot (1–2-m) column of distilled water and collected at the bottom in a pan connected to a precision electronic scale and a timer (figure 5.1). As the larger, heavier particles fall faster, the increase in weight on the scale through time corresponds to the cumulative size distribution of the sediment. Using the fall distance and time, it is easy to calculate the range of fall velocities. Given the water temperature, you can determine the water's viscosity, or "thickness," and density. Knowing grain density (or mineral composition) and shape, you can relate fall velocities and standard grain sizes. Other analytical instruments use lasers to measure the change in water clarity through time after introducing a sediment sample to arrive at similar assessments of grain size or fall velocity.

In general, the newer methods are faster and incorporate instantaneous electronic formatting of the data, allowing a much greater throughput of samples. Also, they require less manipulation of the sample, again enhancing the speed and accuracy of analysis. However, the newer methods lack standardization from laboratory to laboratory, which makes comparisons of results uncertain. Basic data from a settling tube state that the sample has a distribution of fall velocities the same as that of a set of glass spheres with a specific distribution of sizes. Sometimes, determining grain size from fall velocity does yield a better depiction than a sieve analysis of how a sediment grain acts in nature.

Sieve analysis, though, has the advantages of nearly universally accepted standardization and scores of years of comparative data. Therefore, scientists and engineers can interpret the data with a relatively high level of confidence. But working against sieves is that the analysis of each sample takes a lot of time: 15 minutes or more in the shaker, double that if it is necessary to use many screens; a dozen or more individual weighings of the material trapped on each sieve; and data entry—often resulting in tedium. In addition, you must remember just what the sieve is measuring. Think of a sediment particle as a three-dimensional object with its three axes—x, y, and z—each potentially with a different length. The size of a particle passing through any square sieve opening is limited by the axis with the intermediate length, not the greatest. A pencil stub and a brand-new pencil will pass through the same sieve opening, but a wider carpenter's pencil of the same length as either will not pass through the opening.

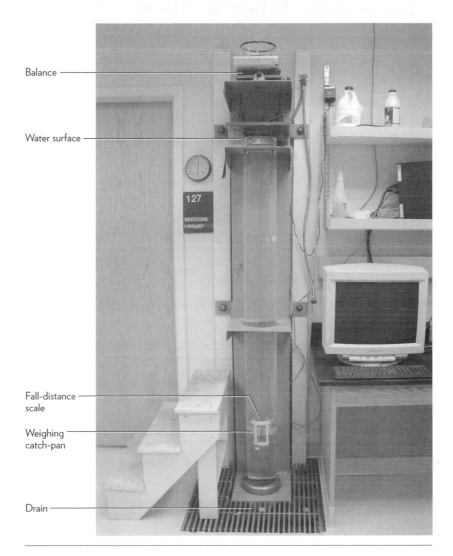

Balance

Water surface

Fall-distance
scale

Weighing
catch-pan

Drain

Figure 5.1 A "settling tube" used to determine the grain-size distribution of sand-size sediments from the sediment fall velocity through water. The weighing catch-pan is suspended from the electronic balance by a thin line. The balance is connected to a computer that both records the data and calculates the grain-size distribution. Introducing the sediment to the water surface with a special mechanism starts a timer. Larger, hence heavier, particles fall more rapidly than and reach the pan before smaller particles. The increase in weight collected on the pan automatically is recorded through time. The grain-size distribution is determined from the distribution of fall velocities, which is calculated from the fall times and the distance between the water surface and the catch-pan, as measured with the tape. After a sample has been run, the sediment can be discarded from the catch-pan by pulling on a small line attached to the edge of the pan. After many samples, the accumulated sediment is removed from the apparatus by opening the drain valve. (Photograph by the author)

Other analytical methods measure the finer-grained sediments: silts and clays. The older, standardized techniques require careful and carefully timed measurements done over many hours for each sample, but several samples may be run in parallel. As with sieve analyses, you must enter the data manually into the computer for plotting, analytical, and archival programs. The modern, computer-assisted methods tend to be quicker, allowing automatic data entry and manipulation, but so far lack standardization and history.

After determining the relative portion of sediment in each small interval, you can calculate several standard statistics. The most commonly used statistics are mean grain size, median grain size, and the sorting or standard deviation of the size distribution. The mean is a measure of the average grain size, while the median is the size at which half by weight of all the particles in the sample are coarser and half are finer. Technical reports often note the median with the term D_{50} (D-fifty). Ideally, the mean and the median should be the same, but nature seldom follows the ideal.

* * *

Why do we care about the grain size or, more properly, the grain-size distribution of sediments? The grain-size statistics, such as the mean and the sorting, provide a way to compare sediments from one place with sediments from other places. Quantitative measures allow us to go beyond saying that one sample is coarser or finer than another, but to be explicit by stating how much coarser or how differently sorted. Also, the grain-size distribution gives clues to the environment in which the natural sediment was deposited. We can use this knowledge to predict how artificially placed sediment, such as construction aggregate or beach fill, will act. Figure 5.2 depicts the results of a laboratory study relating the water velocity necessary to erode and to maintain sediment in transport. In the graph, the deposition field is intuitive; there is a fairly direct relationship between the particle size and the water velocity necessary to keep the particle from being deposited. It takes a current of not quite 1 centimeter per second (cm/s) to keep a 0.1-millimeter-diameter particle from settling and a current of almost 10 cm/s for a 1-millimeter particle.

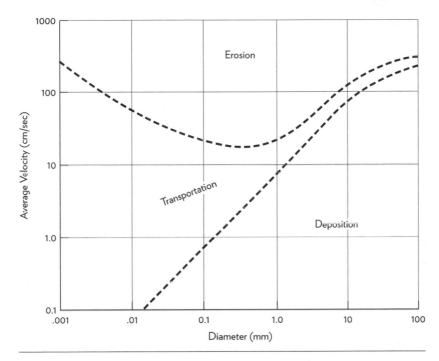

Figure 5.2 The relationship among transportation, deposition, erosion of various sizes of sediments, and current velocity. (Modified from F. Hjulstrom, Transportation of detritus by moving water, in *Recent Marine Sediments: A Symposium*, edited by P. D. Trask [Tulsa, Okla.: American Association of Petroleum Geologists, 1939], in H.-E. Reineck and I. B. Singh, *Depositional Sedimentary Environments, with Reference to Terrigenous Clastics* [New York: Springer-Verlag, 1980]; redrawn by Network Graphics)

Part of the curve for erosion goes against intuition, however. Clearly, it takes a strong current to erode a large particle from the bottom and set it in motion and a lesser current for a smaller particle. But once the particle size is smaller than very fine sand–coarse silt, it becomes more difficult to erode. Or, put another way, the very fine sands are the most easily eroded sediment size class.

One of the reasons that silts and clays resist erosion is their particle shape. Because of their mineralogy and inherent crystal structure, many clay particles tend to be sheet- or leaf-like. Think of pulling apart a "book" of mica and of the many, very small pieces of mica left on your hands.

As with falling leaves, these very fine-grained sediment particles tend to settle with their broad faces toward the bottom as opposed to on edge. This "shingling" significantly hinders the erosion or resuspension of the individual particles. Individual particles lie flat, so there is very, very little relief for the current to push against to lift the particles. Because the particles are deposited in water, the surface tension of the microscopically thin layer of water between the individual leaves helps hold the particles in place. Because the very fine, platy particles resist disturbance, collectively they are considered cohesive sediments, while the more easily agitated, granular sediments are called noncohesive.

The words "clay" and "sand" sometimes are ambiguous, since they convey both size and composition. Sand and clay are grain-size classifications. Clay also is a class of minerals defined by its crystal structure. And, as it happens, clay minerals often constitute the clay-size component of a deposit. Some clay minerals are weathering products of micas. So when using the term "clay," you must specify the context of either size or composition.

"Sand" is a size-classification term. Because most of us are familiar with the white or light-tan beach sands that are composed mainly of particles of quartz and feldspar, however, it is easy to think of sand as just that light-colored stuff and not the full suite of particles. The "black sand" beaches of Hawaii are a well-known example of non-quartz sands. "Dark," "black," or "heavy" sands are common in many areas. They frequently accumulate as patches or lenses on beaches (figure 5.3), often at the inshore edge of the beach against an eroding dune. Dark, or black, sands often are heavy minerals; that is, their densities are appreciably greater than that of quartz. Because of their greater densities, grains of heavy minerals often mix with larger particles of quartz and feldspar, as the grains would have approximately the same weight and, thus, should act similarly.

* * *

So far, the discussion has considered only sediments with a relatively uniform grain size. In actuality, many deposits consist of a mix of sediment types and, not surprisingly, several sets of terms describe the mix.

Figure 5.3 Heavy (dark) minerals on the beach at Cedar Island, Virginia, late August 2006. The lighter-colored minerals primarily are quartz and various feldspars. The dark minerals are dominantly ilmenite, an iron–titanium oxide with a density about 1.8 times greater than that of quartz. (Photograph by the author)

Figure 5.4 presents a common classification that uses the sand:silt:clay ratio of various sediments. When plotting the ratio on a ternary (triangular) diagram, the location indicates the class of sediment. A sample that is 75 percent or more sand or silt or clay is called that type of sediment. A sample that is 100 percent sand plots at the sand apex of the triangle, whereas one that is 0 percent sand plots somewhere, determined by the proportions of silt and clay, along the base of the triangle.

The shape of the grain is yet another property of sediments. The two broad categories of grain shape are roundness and sphericity. Essentially, sphericity is the relative equality of a particle's three axes (up–down, left–right, front–back). A ball and a cube have the same sphericity, since their three axes are of equal length; however, a ball certainly is more round than a cube. Although complicated mathematical methods quan-

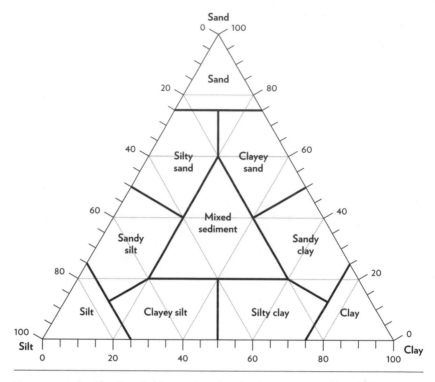

Figure 5.4 A classification of sediments based on the three-way ratio of sand, silt, and clay. (After F. P. Shepard, Nomenclature based on sand-silt-clay ratios: An interim report, *Journal of Sedimentary Petrology* 24 [1954]: 151–158; redrawn by Network Graphics)

tify roundness and sphericity, our eyes are good tools for judging these characteristics. Figure 5.5 provides two charts commonly used in that assessment. The two shape characteristics reflect different things. At its core, sphericity comes from the crystal structure or chemical composition of the grain. Micaceous minerals are very platy; that is, one axis is much shorter than the other two. Quartz and many feldspars often have axes of roughly equal lengths. Zircon tends to be more rod-like. Roundness, however, indicates the grain's transportation history. Particles subjected to many cycles of transport tend to be more rounded because any sharp edges have been abraded. By contrast, grains freshly eroded from a larger rock may have sharp corners. The relative degree of roundness is

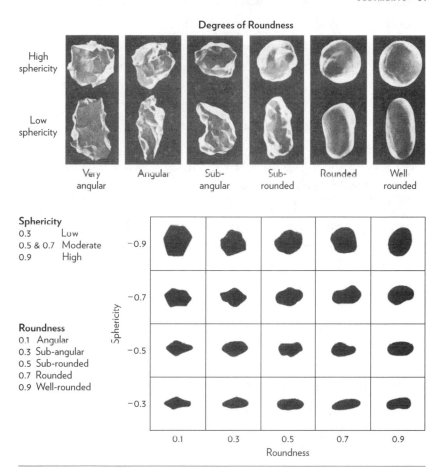

Figure 5.5 Examples of charts used for the visual assessment of roundness and sphericity. (From R. G. Swanson, *Sample Examination Manual* [Tulsa, Okla.: American Association of Petroleum Geologists, 1981]. Copyright © 1981 The American Association of Petroleum Geologists, reprinted by permission of the AAPG. [A] Probably reprinted from M. C. Powers, A new roundness scale for sedimentary particles, *Journal of Sedimentary Petrology* 23 [1953]: 117–119)

one facet of what a geologist would term the "maturity" of the sediments in a deposit.

In addition to suggesting their physical history, grain shape indicates how the particles may act in some circumstances. Grains with uniformly high sphericity and roundness probably would pack together differently from platy, rod-like, or jagged particles. Hence they would have different engineering properties when used, for example, as a substrate for highway construction.

Grain frosting is a surface texture that reveals the physical history of the particle. When observed with a magnifying glass or microscope, some grains appear to be frosted—that is, to exhibit the smooth, dull surface seen on "beach glass." Frosting comes from the abrasion of particles that repeatedly hit and rub against one another, as they do in the surf zone or in areas where there is substantial *aeolian*, or wind-blown, transportation.

Geologists use both the grain-size distributions of individual samples and the relationships between different parameters for suites of samples to learn about the deposits from which the samples were taken. For example, they plot mean grain size versus sorting. Sequential changes in grain-size characteristics within a particular deposit illustrate changes in the environment.

Aeolian deposits tend to be very well sorted—that is, they have a relatively narrow range of grain sizes—but have an asymmetrical grain-size distribution, with an excess of finer particles. Using technical terminology, the distribution has a positive skewness. Individual grains also tend to be relatively well rounded, and many are frosted. The reasons behind these characteristics are easy to understand. Rounding and frosting result from the lack of lubrication during sediment transport. When waterborne particles strike one another during transport, they are cushioned by water and suffer less damage than their air-borne counterparts. Wind damage causes the microscopic pitting that causes frosting, abrasion, and eventual wearing away of sharp corners, thus yielding more roundness. Being less dense and less viscous than water, air is less able to move larger particles than is water, but a steady wind can carry silts great distances. Over the course of many, many years, the wind will strip the finest grains out of a sedimentary deposit while moving the medium-size materials a lesser distance. In an area where the predominant wind blows from a fairly consistent direction, the average grain size will get smaller and the grain-size distribution will become better sorted (more uniform) down-wind.

The absence of sand or coarser particles from a deposit means that the energy regime that formed it was not powerful enough to transport the larger particles. However, if the source area did not have particles of a particular size or mineralogy, grains with those characteristics cannot occur in the deposit.

* * *

The way grains pack together has consequences for the porosity and density of a deposit. The simplest case is packing spheres of uniform size (figure 5.6). Think of filling a box with Ping-Pong balls. The spheres can be neatly stacked in even columns and rows, a pattern called cubic close packing (figure 5.6A). Or they can be neatly arranged with the spheres in one row nested into the hollows between the spheres in the adjacent rows, a pattern called hexagonal close packing because each sphere touches six other spheres (figure 5.6B). Both arrangements work in three

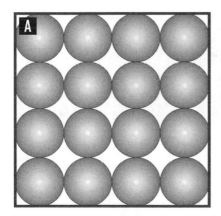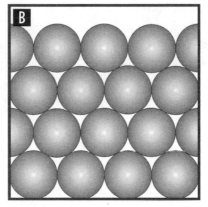

Figure 5.6 The packing of the same number of uniform spheres in two boxes of the same size: (A) cubic close packing, which is inefficient and leaves a large amount of space between the spheres; (B) hexagonal close packing, which is more efficient and leaves both a smaller amount of space between the spheres and room at the top of the box. (Illustration by the author)

dimensions as well as in the two dimensions depicted in the figure. Although the drawing of the hexagonally packed spheres shows some of them cut in half, it demonstrates that a considerably greater number of hexagonally packed balls than cubically packed balls of the same size can be packed in a box. Thus the more closely packed box would weigh more than the other box. When the sizes and shapes of the particles packed in the box vary, small pieces can fit into the voids between large pieces. The mix of different sizes disrupts the potentially neat arrangement allowed by identical sizes and shapes.

The void space between the spheres yields *porosity*, as in the pores between the grains. Porosity is the ratio of the volume of the pore space to the total volume. The volume of the pore space can be measured as the total volume minus the volume of the solid grains. In nature, water often fills the pores between sediment grains not filled by finer-grained sediments. In a seabed or lake bed, water obviously fills the pores. On land, the water that fills the pores is called *groundwater*. Another important property of a body of sediments is *permeability*, or the ability of a porous material to allow a liquid to flow through it. Regardless of the porosity, the voids must connect to one another to allow flow. The flow, or connectivity of the pores, is significant because permeability determines, for example, a well's potential yield. A bed of clean gravel has a much better permeability than a bed of silt and clay with the same porosity. Over the past few years, there has been a lot of discussion in the media about the process known as "fracking" (hydraulic fracturing), initiating and spreading a fracture in rock—for example, a shale bed—to enhance the production of natural gas. Fracking works by increasing the permeability of the bed, so the gas can flow through it to the extraction well.

A commonly asked but impossible-to-answer question goes something like: How many grains of sand are there on this beach? Think about spherical grains inefficiently arranged in cubic close packing, with grains of 0.5-millimeter diameter (the separation between coarse and medium sand). If we line up the particles in a neat row, two of them will make a line 1 millimeter long, and 20 will span 1 centimeter. So 1 cubic centimeter will contain 8,000 (20 × 20 × 20) poorly packed grains of this sand. Thus a 2-liter (2,000 cm³) soda bottle will contain 16 million grains. However, if we use smaller particles—say, 0.25 millimeter (so a 1-centimeter line contains 40 particles)—the number of grains in the soda bottle jumps to 128 million! Remember that this is a minimum number; in real life, even if we did have perfectly spherical particles of the designated size, they would pack together more efficiently, so it would take more grains to fill the bottle. The number of particles in a dump truck or on a beach is unimaginably huge.

* * *

Finally, your child may ask, "Where does sand come from?" The simple answer is that sand is a product of the mechanical and chemical weathering of rocks. Whenever rocks crash into other rocks, expand or contract with heating and cooling, or experience the pressure of water freezing and expanding in crevasses or pores, the rocks fracture into smaller particles. Given enough time, water dissolves many minerals. Most rocks are a combination of different minerals. As water dissolves some of the minerals in a rock, some grains of the remaining minerals fall out of the rock as particles of sand. When these processes continue for an extremely long time, the result is a tremendous quantity of sand. Because particles of similar sizes and densities respond similarly when transported, they tend to be deposited together.

Sometimes it is possible to be more specific about stating where sand comes from. Specific minerals or trace elements within minerals that occur in only one place in a drainage basin indicate that the sediments making up the deposit, at least in part, came from that particular area. This place is called the *provenance* of the sediment. Careful mineralogical and geochemical analyses can match individual sediments and potential source areas. Small differences in the chemical composition of individual minerals or in the isotopes of some elements can lead to source areas.

* * *

Sediments are the material of the beach. Individual particles are described by their size, shape, and composition. Their characteristics influence when they are eroded and how they are deposited. With knowledge of the grain-size distribution of a sedimentary deposit, we can develop an understanding of a physical environment at the time and place of deposition. Other aspects of sediments, such as whether particles are frosted like beach glass, may provide further detail and the geological history of a deposit. The way sediments pack together along with the variation in grain sizes within a deposit are major determinants in the porosity and permeability of the deposit, and these two parameters control how much fluid—whether water, oil, or natural gas—can be held in the deposit and how difficult or easy it is to pump the fluid to the surface. Thus the characteristics of sediments have implications for such contentious issues as oil and gas drilling.

6
BARRIER ISLANDS AND TIDAL INLETS

Barrier spits and barrier islands buffer the mainland shore from the wave energy of the ocean. The origin of barrier spits is clear even to unsophisticated observers. A strong, net longshore current transports sediments from a source area, often a promontory. As the sediments move with the longshore current, the spit grows longer. Wave refraction and diffraction wrap around the end of the spit, creating a hook, or "re-curve," as some of the sediments and the spit turn the corner into the back-barrier bay.

The sediments along a spit show the consequences of the transportation processes. Sediment particles that are more easily eroded from their bed and more easily kept in motion will, as a class, travel farther. Because finer particles stay suspended in water longer than larger particles, they move down-drift faster than coarser materials. Thus if the sediments at the source area contain a mix of grain sizes, they tend to become progressively finer downstream along the spit. If a barrier spit grows long enough, it can break, or be breached, forming a barrier island. However, this is not the only way in which barrier islands can form.

Another major theory for the genesis of barrier islands is depicted in figure 6.1. According to this generally accepted view, there are five requirements for the formation and existence of barrier islands:

- A gently sloping mainland surface
- A rising sea level

- Energetic waves
- A (generous) supply of sand
- A low to intermediate tidal range

In stage I, there is an open coast with a beach-and-dune system between rivers. The region behind the beach is a gently sloping coastal plain. As sea level rises in stage II, the rivers widen to become estuaries; low-lying areas behind the higher dunes are inundated, and marshes grow. Finally, in stage III, the low-lying areas landward of the dunes are fully flooded,

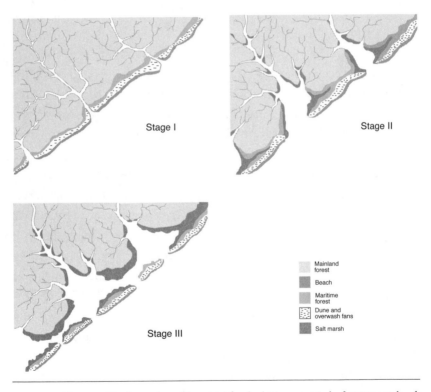

Figure 6.1 The possible formation of barrier islands during a period of rising sea level: beginning with beaches on a low-lying mainland (stage I), the rising water floods the rivers that drain into the sea, widening the inlets between the beaches (stage II); a further rise in sea level inundates the area immediately behind the beaches, separating them from the mainland (stage III). (From O. Pilkey, *A Celebration of the World's Barrier Islands* [New York: Columbia University Press, 2003], 36)

separating the beach–dune complex from the mainland. If the mainland were high ground, it could not flood and be separated from the beach. If it were steep instead of gently sloping, the flooded area would be very narrow, probably insufficient to allow the development of the back-barrier lagoon. If relative sea level were not rising, the flooding could not occur; if it were rising too rapidly, the water might overtop and submerge the beach. In an area without waves, it is unlikely that a sandy beach would develop. Similarly, without sand, the beach-dune system could not come into being. Finally, as will be considered later, tidal range plays an important role in the evolution of barrier islands.

Yet another theory for the origin of barrier islands is that they begin as sandbars that have grown through a varied set of processes until they remain above water level. Portions of Dawson Shoal, the ebb delta at Wachapreague Inlet, Virginia, have remained above water for most of the past forty or more years. Although waves wash the ebb delta during severe storms and it occasionally has disappeared as an island, it has been emergent for enough time that plants grow on it and it has been large enough for an airplane to land on it.

Tidal inlets, the breaks between barrier islands, exhibit a strong relationship between tidal range and inlet size. When the tidal range is large, inlets must be large enough to handle the volume of water that has to flow through them to flood, or drain, the back barrier. At some point, the inlets become so large relative to the size of the barrier islands and the tidal currents so great that the islands cease to exist as nearly shore-parallel features. They become more elongate in the cross-shore direction such that they cease to be barrier islands.

Barrier islands do not develop in regions of high tidal range, but the influence of tidal range on barrier islands is more nuanced than the comment suggests. There is a continuum according to which the higher the tidal range, the shorter the islands in a chain, as seen along the southern Atlantic coast of the United States. In North Carolina, the tidal range is relatively low, with mean ranges of 3.4 and 3.7 feet (1 and 1.1 m) at Capes Hatteras and Lookout, respectively. In South Carolina, the range increases: 5.1 feet (1.6 m) at Garden City; 5.8 feet (1.8 m) at Edisto Beach; and 6.7 feet (2 m) at Hilton Head. At the Egg Islands, off the coast of southern Georgia, the tidal range is 7.2 feet (2.2 m). A quick scan of coastal maps and charts suggests that the barrier islands in North Carolina are longer than those farther south. The longest unbroken barrier island in the United States is Padre Island, off the Texas coast of the Gulf of Mexico, where the tidal range is less than 1.5 feet (0.5 m).

While spits have a direct connection to a sediment source, barrier islands, by their very nature, are detached from that immediate source and thus may be more vulnerable than spits to sand loss. Barrier islands rely on natural sand reservoirs for their existence, although they may change through time. The most obvious natural sand reservoirs are the dune system and the wide berm and back beach.

* * *

Dunes are an integral part of barrier beaches. Given a growing beach and a supply of sand, wind blows the dry sand. Aeolian transport is a sorting mechanism. Wind cannot move large particles of very coarse sands, while it can carry very fine sands and silts for great distances. The sediments in between—fine and medium sands—get buffeted back and forth and around and around, such that they collect, pile up, and form dunes. Because wind-blown particles move at higher velocities than water-borne grains and because the particle-to-particle impacts in aeolian transport are not buffered or lubricated by water, the collisions between grains are harder and abrade the grains. Viewed under a magnifying glass, dune sands appear to be frosted, as is beach glass. Thus the occurrence, in what geologists call the "rock record," of very well-sorted (as to grain size), finer-grained sands with frosted surfaces generally is diagnostic of deposits of wind-blown sediment.

Even as a mini-dune starts to form, it grows larger as sediment falls into the dune's wind shadow. This sediment accumulation in a lee area is obvious where waves cut the dune during a storm and a distinct pile of sand builds at the foot of this steep face. As the wind often shifts from onshore to offshore as the storm passes, the strong offshore wind blows a lot of sand back onto the beach. The sudden drop in wind speed in the lee of the dune causes the wind-borne sediment to fall.

Dune vegetation like sea oats (*Uniola paniculata*) and American beach grass (*Ammophilia breviligulata*) help stabilize the dunes. Wind cannot denude the vegetated sand nearly as well as it can scour a bare surface. Plants that grow in the dunes must be exceptionally tolerant of salt, heat, and aridity. The salt spray that dune grasses endure kills many other plants. Because the light-colored sands reflect a lot of solar energy and there is little shade, the plants literally get cooked for days on end during the summer. And as sands tend to be very permeable—that is, they do not hold water—the plants must develop long root systems to capture water. The deep-reaching roots help bind the sand, further stabilizing the dunes.

When storm waves surge over a barrier island, the process of *overwash* simultaneously destroys and preserves the island. Overwash often is channeled through low spots, sometimes called *blowouts*, in the dunes. Although the term "blowout" suggests that blowing wind created the low passage through the dune field, the process is more complicated. Undoubtedly, winds do help maintain the low spot. More likely, the blowout formed when storm flooding breached an existing low or weak area in the dune front.

The strong, turbulent flow of storm waters can deepen and widen the blowout. Waves push the sand removed from the dune front through the blowout and deposit it in a delta-like washover fan on the back side of the island (figure 6.2). A washover fan can change the local environment

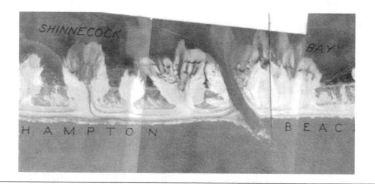

Figure 6.2 A mosaic of aerial photographs of Shinnecock Inlet on the south shore of Long Island, New York, clearly depicts the major washover fans and the new inlet formed by the Great Hurricane of 1938. (From U.S. Army Corps of Engineers, U.S. Army Engineer Research and Development Center, Coastal Inlets Research Program, Inlets Online, http://www.oceanscience.net/inletsonline)

by covering a marsh surface with a sand sheet. More often, sand is spread across the marsh, but a lot of the marsh vegetation still protrudes above the surface. The vegetation makes it difficult for the wind to carry the sand back toward the beach.

By eroding the dune front and moving large quantities of sand from the berm and beach face, overwash appears to deplete the beach. But by moving sand from front to back, overwash is the significant process of island rollover or migration. By migrating landward and upward, barrier islands can accommodate sea-level rise. Given a rising sea level, without an influx of sediment, barrier islands are doomed unless they redistribute their sediments and continually rebuild themselves. Figure 6.3 shows barrier-island rollover on Cedar Island, Virginia. As the dunes move over the back parts of the island, they engulf trees and shrubs in the relatively protected back-barrier region (figure 6.3A) and even shift entirely onto a marsh, so marsh deposits that once were behind the island are exposed on the beach face (figure 6.3B). And barrier-island rollover has human and economic consequences (figure 6.3C). The house had been built on the dunes, and at least part of the long pilings penetrated through the dunes into the underlying sediments. As the island rolled landward, the now uninhabitable house stayed in place above the low-tide portion of the beach. These three photographs were taken in three areas of Cedar Island within a minute's walk of one another.

Parramore Island, Virginia, is another example of rollover processes in action. Figure 6.4 shows part of a maritime forest exposed on the beach and in the surf. Recently, the shoreline has retreated approximately 50 feet (15 m) a year. This rapid erosion likely will continue until the shoreline reaches higher ground.

In the absence of washovers along substantial portions of the Outer Banks of North Carolina where the dunes are large and unbroken, beach-front sediment is lost to both offshore transport and longshore drift. Storms may carry the sand beyond the fair-weather closure depth. Where the dunes are strong and unbroken, the islands erode on the front and flood on both the front and the back by the rising ocean, so the island narrows through time. Thus the dune system is a paradox. Its huge reservoir of sand protects the back-barrier areas from the ravages of storms. At the

Figure 6.3 An example of barrier-island rollover on Cedar Island, Virginia, September 2007: (A) sand dunes have moved landward into the trees that had grown in a previously sheltered, back barrier region; (B) a marsh that originally formed between the barrier island and the mainland is now exposed on the beach face; (C) as the sand dunes and beach face retreated toward land, a house that had been built in the dunes was left standing above the low-tide portion of the beach. (Photographs by the author)

Figure 6.4 The ocean shoreline of Parramore Island, Virginia, is retreating into the well-established maritime forest. The recent rate of erosion is approximately 50 feet (15 m) a year. (Photograph by the author)

same time, it might be accelerating the demise of the island by blocking a natural response to rising sea level. This dual process is especially interesting along parts of the national seashore in North Carolina, where the dunes are a major feature. In the 1930s, the Civilian Conservation Corps and the Works Progress Administration built the dunes in an attempt to protect the islands while providing work to people who had lost their jobs during the Great Depression.

That said, dunes, especially naturally occurring dunes, are fragile. Simply hiking a few times along the same path through a dune field can kill the beach grasses and encourage the development of a blowout in a place where one otherwise might not have formed. Driving through the dunes can do the same thing, only more quickly. Although imperfect, walkways elevated on pilings above the dunes offer some protection to the fragile ecosystem.

* * *

Tidal inlets are an integral part of any discussion of barrier-island dynamics. Not counting inlets that are river mouths, such as the Merrimack River Inlet between Plum Island and Salisbury Beach in northern Massachusetts, tidal inlets connect back-barrier bays or lagoons to the open ocean. Inlets interrupt the movement of sediment along the shoreline. Sediment moving down the barrier island eventually reaches the inlet at the down-drift end of the island and may get caught in the tidal currents that flow through the inlet.

The geomorphology of tidal inlets is complex. Careful observations can provide useful lessons in the workings of an inlet. Figure 6.5 depicts the major features of an inlet. The marginal flood channels are conduits for a

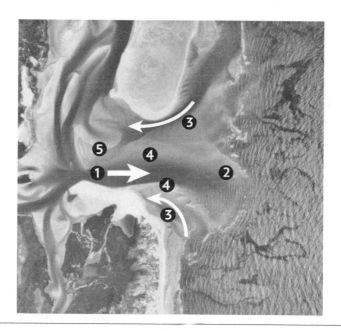

Figure 6.5 The elements of a tidal inlet–tidal delta system: (1) main ebb channel; (2) ebb delta terminal lobe; (3) marginal flood channel; (4) channel margin bar; (5) flood delta platform. The dominant wave approach is from the top right. (Aerial photograph from U.S. Army Corps of Engineers, U.S. Army Engineer Research and Development Center, Coastal Inlets Research Program, Inlets Online, http://www.oceanscience.net/inletsonline)

lot of water and sediment. After low water, as the tide begins to rise, the flow cannot utilize the main channel because the water level is too low to move across the ebb lobe. As a result, early in the flood, the currents are forced around the lobe, where they first form and then use and maintain the marginal flood channels. The tidal current flooding into the inlet reinforces the background, wave-generated, longshore current. When the dominant waves from the northeast cause a south-flowing current, as seen in figure 6.5, the effect is most noticeable at the northern side of the inlet. The combined flood and longshore currents carry sediments that hook into the inlet, forming small spits on the distal end of the barrier island that eventually lengthen the island. In order to maintain the inlet's constant cross-sectional area, however, the growth of the northern barrier island is balanced by erosion at the tip of the southern barrier island, on the other side of the inlet. In effect, the inlet migrates in the direction of the dominant longshore current.

Not all inlets migrate. At some tidal inlets, another factor, such as the underlying geology, serves to fix the location. The hydrodynamic forces work to clean freshly imported sediment from the inlet and move it to the main deposits of the ebb and flood deltas.

Storms can use a breach in the dunes to create a new inlet. This is apt to happen if both storm surge and substantial waves flow through a breach in the dunes in the same process that forms a washover fan. When the wind shifts as the storm passes, if the surge is high enough, wind-driven water escapes from the flooded back-barrier bay through the new opening. Then, if the new inlet can capture a large portion of the normal tidal flow, it can become dominant, and the old inlet may close. This process has happened innumerable times through history, and not always where property owners want a new inlet. Looking at topographic maps or aerial photographs of barrier islands, you can locate the probable sites of the flood deltas of abandoned inlets. They are marshy bulges on the back of the islands.

Given society's investment in physical continuity, however, humans often step in and attempt to "repair" the damage by closing the new inlet to maintain the old. This occurred at Cape Hatteras, North Carolina, following Hurricane Isabel's landfall on September 18, 2003. The Category 2 hurricane created a breach 1,700 feet (500 m) wide and a new inlet (figure 6.6),

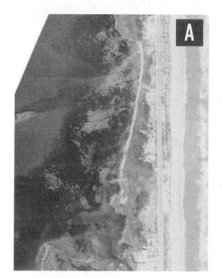

Figure 6.6 The inlet north of Cape Hatteras, North Carolina, formed during the landfall of Hurricane Isabel, on September 18, 2003: (A) Cape Hatteras in 1998; (B) the same area on September 19, 2003; (C) an oblique photograph, taken a few days later, shows the development of significant inlet channels and the beginning of the transformation of the washover fan into the flood delta. ([A] and [B] National Oceanographic and Atmospheric Administration, http://www.noaanews.noaa.gov/stories/images/hatteras-new-inlet-comparison .jpg; [C] North Carolina Coastal Federation, Cape Hatteras Coast Keeper issues: Hurricane Isabel damage report, http://www.nccoast.org/ch-issues.htm)

with depths in the channel reaching 20 or 30 feet (6–9 m) and currents of 6 to 7 knots. The new inlet separated the village of Hatteras from the rest of the island to the north. Before the Army Corps of Engineers acted to close the inlet, there was speculation that it might become a new "permanent" inlet. The same thing happened again on August 27, 2011. Hurricane Irene breached the barrier islands of North Carolina's Outer Banks in four places, severing the highway and isolating the village of Hatteras. Each side has merit in the debate about the socioeconomic need for control versus interference with natural processes.

Society's desire to control the natural processes associated with tidal inlets is evident in the engineering projects associated with many inlets. These usually are attempts to "stabilize" the inlet by building jetties on one or both sides or by dredging to maintain and better define a channel for navigation. The economic impetus to control an inlet makes sense. Open-ocean coasts provide no protection from waves and storms, so they are unsuitable maritime ports. Inside the back-barrier bay or lagoon, there is protection from ocean waves. And we demand navigation access: recreational anglers and boaters, open-ocean scuba divers, commercial-fishing operations, and commercial and military shipping all benefit from direct access to the ocean from sheltered harbors. The dollar costs of the engineering projects have to be balanced against both the economic gains and the environmental losses of altering the natural systems. It is difficult to establish a monetary value for many environmental consequences, so they often do not receive the consideration that they deserve.

The basic morphology, or shape, of tidal inlets makes navigation through them difficult. The marginal flood channels are shallow and can have strong currents. The ebb lobe naturally blocks access to the main ebb channel. In rough weather, waves break on the ebb lobe. The low spots on the ebb lobe and the shoals between the lobe and the main channel frequently shift. Then, if the inlet is trying to migrate, sediment from the up-drift regions further complicates the shifting channels. Combining all these hindrances to navigation, it is logical to want to control inlets, to extend the main ebb channel completely through the ebb lobe and into relatively deep water, and to stabilize the various interior channels.

Jetties are engineering structures built to enhance navigation at inlets. Jetties should not be confused with *groins* (or groynes), which are erosion-control structures. Jetties sometimes are constructed singly, on the up-drift side of an inlet, or in pairs with one on either side of the inlet. They have three general functions:

1. They impede or halt the transport of sand into the inlet with the general longshore drift (in which case, they do act as groins).
2. By extending into deeper water, they assist navigation "across the bar."
3. Finally, they fix the location of the channel, which also benefits navigation.

The engineering challenge is substantial. If spaced properly, jetties define a narrow channel, so natural processes keep it open. Given the dynamic between the tidal prism and the inlet cross section, a narrow channel will cause the flow to jet (hence the term "jetty") through the opening and scour the channel bottom. By extending across the ebb lobe, the jet can disperse the sediment in deeper water.

Unfortunately, controlling nature rarely is that easy. Because the up-drift jetty does trap sediment like a groin, its construction has significant consequences: the beach at the up-drift jetty builds out, often to the end of the jetty, and the shore across the inlet erodes because of the disruption of the natural flow of sediment. Figure 6.7 depicts the landward movement of Assateague Island immediately down-drift of the inlet at Ocean City, Maryland, over 31 years. Thus the construction of a jetty will require continued intervention with nature for a long time.

Tidal deltas do not just spring into existence fully formed. If left alone, an ebb delta can grow sufficiently large that it serves as a bridge in the longshore system carrying sediment fully across the inlet. But ebb deltas in developed areas rarely are left alone. The potentially large volume of sand comprising the delta is an obvious target for exploitation.

An ebb delta also affects the down-drift island and beach. By its very nature, an ebb delta results in shallower water farther offshore. Thus waves "feel the bottom" farther offshore and begin to refract farther offshore. Loosely put, the waves "wrap" around the outer portions of the ebb delta and locally perturb the wave-driven, longshore sediment trans-

Figure 6.7 The inlet at Ocean City, Maryland: (A) the inlet in September 1933, after it opened during the hurricane of August 23, before the construction of the jetties; the beach south of the inlet, Assateague Island, essentially lined up with the beach north of the inlet; (B) the inlet in May 1964, after the famous Ash Wednesday Storm of March 6–8, 1962, during which much of Assateague Island experienced overwash from the 6-foot (2-m) storm surge and high waves. In the 31 years between the two storms, the northern end of Assateague Island moved landward by the width of the island. (From U.S. Army Corps of Engineers, U.S. Army Engineer Research and Development Center, Coastal Inlets Research Program, Inlets Online, http://www.oceanscience.net/inletsonline)

port. Along much of the East Coast of the United States, the important waves approach the shore from the northeast, driving sediment to the south. But as the waves reach an ebb delta, two reinforcing phenomena happen:

1. The delta shields the area to its southwest from the full force of the waves.
2. Perhaps more important, the waves that refract around the outer lobes of the delta change direction, refract, and locally approach the shore from the southeast, driving sediment back toward the inlet.

The accumulation of sediment at the northern end of the barrier island pushes the shoreline seaward so that the opposite sides of the inlet can be offset from each other. The down-drift side of the inlet extends farther out to sea.

The stretch of beach where the longshore drift reverses can experience long-term erosion greater than that at adjacent shorelines. Sand removed from this zone, or "nodal" region, is not replaced. The longshore current down-drift of the nodal region and the reversed upstream current carry sediment away. So the nodal region is scoured, and the barrier island becomes narrower.

Some barrier islands look like turkey drumsticks: fat at the top end, skinny in the middle, and slightly rounded at the bottom (figure 6.8). The

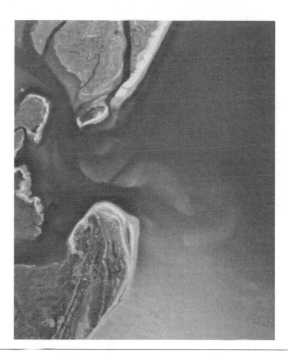

Figure 6.8 An infrared photograph of Quinby Inlet, on the lower Delmarva Peninsula, Virginia, depicts an offset inlet and an ebb-tidal delta. The growth of the bulbous end of Hog Island (the southern island) is evident in the small spits in the beach extending toward the inlet. (U.S. Geological Survey, Quinby Inlet NE DOQQ [Digital Orthophoto Quarter Quadrangle (infrared)], http://iris.lib.virginia.edu/mrsid/bin/get_image.pl?image=/lv5/siddoq/37075/37075d66.sid&size=thumbnail)

bulbous, up-drift end grows seaward, and a feathery distal end wraps into the down-drift tidal inlet. As well as creating an offset inlet, this process causes the island to rotate. The fat, up-drift end of the island grows outward while the thin, down-drift end curls into the inlet.

* * *

Beaches are an integral feature of barrier islands. For many people, the idea of "going to the beach" means going to a barrier island. The houses, hotels, restaurants, and other amusements are as important to some visitors as the sandy beach berm and the surf. Tidal inlets are important, too. They provide access to harbors and calm water. If we understand how nature builds and changes beaches, barrier islands, and tidal inlets, we can make better decisions about what we do to them and how we live with them.

7
SAND DUNES AND SALT MARSHES

Sand dunes and salt marshes are common features near beaches. Sand dunes occur where there is sufficient sand and enough wind to move it. They are an aeolian deposit, meaning that they develop from wind-blown sediments. In Greek mythology, Aeolus is the god of the winds. According to the Army Corps of Engineers' *Shore Protection Manual*, "Winds blowing inland over the foreshore and berm move sand behind the beach to form dunes. Grass, and sometime bushes and trees, grow on the dunes, and the dunes become natural levees against sea attack. Dunes are the final natural protection line against wave attack, and are also a reservoir for storage of sand against storm waves." Sand dunes can form inland as well as at the coast. The dunes of the Sahara Desert of Africa and the Arabian Peninsula are among the examples that usually come to mind. The difference between these and coastal dunes is the presence of water and vegetation along the shore.

Sand dunes have an important function in the beach system. A relatively small storm may cut the seaward side of a dune, leaving a steep face, or *scarp*. All the sand liberated from the dune then is transferred to the active beach. In a relatively strong storm, however, the waves, abetted by the elevated water level, may breach the dune. The resulting gaps in the dune facilitate washover. The flowing water not only can transport sand to the back side of the dune, but also wash away the sides of the dune.

Once eroded, dunes may rebuild, but they do not return to their original structure. Over time, however, new deposition may enlarge the remnant almost to the size of the original dunes.

* * *

Wind moves sand in a surprisingly complex manner, with consequences on several physical scales. Because air is much, much less dense than water, the force applied by wind is less than that applied by water moving at the same speed. The combination of low density and weak force limits the wind's ability to move sediments. Because larger particles are heavier than smaller particles of the same composition, they are less likely to be transported by wind. So, simplistically, the sediments in aeolian deposits potentially are better sorted than those in other depositional environments. Well-sorted sediments have a relatively smaller range of grain sizes than poorly sorted deposits. And, just as in water, finer sediments have lower fall velocities than larger particles. Therefore, the finer sands are more likely to remain airborne after the coarser sands have fallen to the ground.

In a hypothetical area originally covered by poorly sorted sand and subject to a constant wind from one direction, the sediments likely would be spread out as time passes. The wind would leave the coarser particles behind and carry the finest particles farthest. Over time, this differentiation could be measured: mean grain size would decrease, and sorting would increase down-wind. Geologists often observe this phenomenon on long barrier spits or islands where wind from one direction is dominant. Also, just as with water-borne sediments, mineralogy is important in aeolian transport. Dense particles settle out of the air flow faster than less dense particles of the same size. Therefore, the dark (heavy) minerals concentrate in areas where the wind speed decreases.

The abrasion from wind-blown sand that stings our legs also changes the sediments. Think of the pieces of beach glass that many people collect. The impacts of thousands of sand grains on smooth glass microscopically pit the surface, leaving the cloudy matte finish that distinguishes beach glass. The same abrasion happens to individual sand grains. The pitting imparts a frosted finish to the grains that is diagnostic of aeolian sediments. This natural sand-blasting also abrades rocks, wooden pilings, painted structures, and almost anything else it hits. *Ventifacts* are rocks that have been polished or otherwise sculpted by aeolian processes.

The wind-blown sand usually stings only our legs because the wind cannot lift sand far above the ground. Although the horizontal velocity of the wind is fast enough to pick up and carry a lot of sand, the wind's upward component is too low to overcome the sand's downward velocity. Therefore, even if initially lifted high above the ground, the grains cannot remain far aloft. Near the ground, where the wind is subject to bottom friction, there is more turbulence than where the flow is undisturbed, and there are greater upward flows. Thus the wind-blown sediments may concentrate in a relatively thin zone immediately above the surface. The materials transported in sand storms or dust storms tend to be relatively fine grained. Stronger winds can lift the particles higher, and finer grains, with their lower fall velocities, go higher still. The monumental dust storms, sometimes called *haboobs*, that we see in the deserts of the Middle East and that tormented the Midwest of the United States during the 1930s and 1940s and engulfed Phoenix, Arizona, in the summer of 2011 generally are composed of very fine-grained sediments.

Over time, wind can move vast quantities of sediment. Sand dunes can be hundreds of feet tall, and some linear dunes extend for many miles. The *loess* deposits just east of the Mississippi River in parts of Tennessee and Mississippi are 100 feet (30 m) thick.

Sediments move in three general classes of transport: suspension, saltation, and traction. Finer particles stay aloft because the speed of the vertical air current exceeds the fall velocities of the particles; this is the suspended load of the flowing current. Grains that move by saltation bounce along near the bottom. They move at a horizontal speed that approaches that of the current, whether it be air or water. Motion usually begins when one grain hits another grain at rest. The momentum, or force, of the moving particle that strikes a grain on the ground transfers to that particle. Remember the $F = mv^2$ calculations; if the force is great enough, it will kick the particle from the ground up into the flow. And because the force increases with the square of the velocity and, especially in storms, the wind's velocity can be great, the force of a moving grain of sand can easily be enough to dislodge the grains it strikes. As it moves into the current, other particles may strike it and push it even higher. Gravity wins in the end, and the particle begins to fall while maintaining a horizontal speed very close to that of the ambient current. When the particle hits the bottom, the process starts again.

Particles that move by traction tend to be larger than those transported by suspension and saltation. They move when other particles hit them and the current pushes against their sides. The traction load of a flow moves along the bottom and does not really rise above it. The particles move over other particles on the bed, but generally stay in contact with one another. Together, the sediments transported by saltation and traction are called the *bed load* of the flow. These processes apply equally to wind- and water-borne sediments. Because water is much denser than air and has a much greater viscosity, the scales of the forces and the trajectories of the particles differ, but the modes of movement are the same in both air and water.

* * *

The direction(s) from which the stronger winds blow is important in the development of sand dunes. If the dominant wind flows from the land, it carries sand into the water, where it, temporarily at least, leaves the dry portion of the beach and joins the longshore transport system. If the dominant wind blows from the water, it can dry the sand brought onto the beach by waves and, later, move that sand to the high, back areas of the beach and beyond where dunes can form. *Foredunes*, sometimes called *primary dunes*, are the first line of dunes landward of the beach. If the area is wide enough and if there is enough sand, secondary dunes can develop farther from the ocean behind the foredunes.

Sand dunes have a variety of shapes. Those that form in regions of generally unidirectional winds are called *transverse dunes*, *barchans*, and *parabolic dunes*. In cross section, a transverse dune looks a lot like a current ripple or sand wave, created by flowing water. It has a relatively shallow up-wind side and a steeper down-wind side. The dune grows and migrates as sand is pushed over the crest of the ridge and falls down-slope and down-wind. In unconsolidated dry sand, the *slipface* can be no steeper than about 32 degrees—the *angle of repose*. Once the sediment packs more tightly or becomes partially cemented, sand faces can attain a steeper angle.

Both barchans and parabolic dunes develop from transverse dunes. A barchan is crescent-shaped, with the arms extending downwind. The reverse shape, with the arms pointing into the wind, is a parabolic dune. Given the unidirectional wind, why do the contrasting shapes develop? The difference lies in the presence or absence of plants. Sand is less easily blown away from vegetated than from bare areas because plants interfere with the wind's ability to couple with the sand surface. Along the higher center portion of a transverse dune, conditions are poorer for vegetation than on the sides. The wind has unhindered access to the sand surface, so the bare center portion of a transverse dune moves downwind faster than the sides, which are anchored by plants. This more rapid advance of the center creates the parabolic shape. A barchan develops from a bare transverse dune because the absence of vegetation allows the relatively smaller quantity of sand on the sides of the dune to move more rapidly than the sand in the more massive center portion. Thus the arms of a barchan ahead of the central body.

Medanos are hills of sand. They form where strong winds blow from multiple directions. Lacking a dominant wind to force the shape of the dune, the sand tends to mound. With little original relief in the surface, aeolian sand tends to accumulate in the lee of an obstruction. If the wind blows first from one direction and then from another, the orientation of the "lee" changes, and the "sand shadow" forms on different faces of the medano.

The wind's selective sorting can erode as well as deposit sand. When the wind blows over a broad, usually dry, sandy area such as the high, landward-most portion of a berm, it removes the finer particles from the sediment surface and leaves the coarser fragments. Over an extended time, the surface becomes a coarse-grained, flat area that sometimes is referred to as a *pavement* (figure 7.1). Shells often comprise the surface of a wind-blown pavement.

Figure 7.1 A wind-blown pavement of shell between low, vegetated dunes on Cedar Island, Virginia, September 2008. (Photograph by the author)

* * *

Plants help sand accumulate. The stems and leaves not only shield the sediment surface, hindering the wind from eroding particles, but disrupt the wind current so sand grains fall either between the shoots of the plant or as a small sand shadow immediately in the lee of larger plants (figure 7.2). It takes several circumstances coming together for plants to start growing. Some organic matter has to be mixed into the sand to form even a minimal soil and to provide nutrients to support plant growth. The organic material can be fragments of seaweed or similar detritus that spring tides or storms have carried to the high portions of the beach as well as bird and insect droppings and the like.

The back beach and dunes are especially harsh environments for plants. The porous and permeable sand usually does not hold water. To

Figure 7.2 Wind-blown sand accumulating in the lee of detritus and new growth high on the berm. (Photograph by the author)

survive in these settings, plants must use rainwater and dew efficiently, and their roots must penetrate far below the surface to approach the water table. The plants also must tolerate salt, from both wind-driven spray and infiltration of ocean water into the groundwater system.

The sand surface itself is a hostile environment. The sunlight that reflects from the surface bakes plants. Anyone who has walked across a wide beach on a bright, hot summer day, especially barefoot, has experienced its harshness. The plants also are at risk of getting buried. As already noted, plants enhance the deposition and trapping of wind-blown sands. The plants that pioneer on high beaches have to be quite specialized.

In sand dunes along the Atlantic coast, American beach grass (*Ammophilia breviligulata*) is common north from Virginia, whereas sea oats (*Uniola paniculata*) grows from Virginia southward. The transition is gradual and the two plants coexist in many areas. Both effectively hold sand and stabilize dunes. And as plant communities grow, they both trap organic matter and contribute it to the sediment, improving the substrate for other plants. The growth of more plants speeds soil development and, perhaps more important, provides an increasing amount of shade for the land surface. With slightly better soil and a somewhat moderated local environment, a wider range of plant species can grow. As plant cover increases and the dunes stabilize, the plant community matures.

Not all coastal sand dunes are natural. In many areas, some sort of mechanical method has created or enhanced sand dunes. Bulldozing; erecting sand fences, which are similar to the drift fences used to catch snow in northern areas; and planting dune grasses are common actions. In the 1930s, the Civilian Conservation Corps (CCC) and the Works Progress Administration (WPA) built lines of brush as sand fences to trap sand and enhance the natural dunes along portions of the Outer Banks of North Carolina. Later, the National Park Service maintained and stabilized the dunes. These efforts successfully developed the sand dunes, but the overall impact on the barrier islands' dynamics may have been negative, as massive dunes can preclude overwash. Without overwash, the island width decreases as the ocean side continues to erode and rising water encroaches from both sides.

* * *

Plants define tidal (salt) marshes, another environment often found near beaches and barrier islands. Less common are tidal, freshwater marshes and mangrove swamps. Tidal, freshwater marshes occur in the upstream tidal portions of some coastal rivers. Black mangrove (*Avicennia germinans*) is limited in the United States to portions of Florida, Louisiana, and Texas. Plants in more temperate tidal marshes must tolerate both frequent inundation and saline water. Plant species vary in their ability to tolerate submergence, so different shore plants live in different zones from the water to the upland. Sea grasses, such as eelgrass (*Zostera marina*) and shoal grass or shoal weed (*Halodule wrightii*), can live underwater.

The tide determines how often and how long the plants of a tidal marsh are submerged. If we think of a tidal shore and impose various tidal datums on a simple cross section, the relative time and frequency of submergence becomes more clear. The region below mean lower low water virtually always is underwater and emerges only when conditions, usually strong offshore wind and high barometric pressure, produce unusually low tides. The portion of the shore between mean lower low water and *mean tide level* (MTL) spends about one-half of the time submerged. Up to the level of mean high water at neap tides (*mean high water neaps* [MHWN]), the shore is likely to be underwater at least once a day in regions with diurnal tides and twice a day in those with semi-diurnal tides. The frequency and duration of submergence decreases with elevation. Around the level of mean high water at spring tides (*mean high water springs* [MHWS]), submergence usually occurs only a few days a month. Higher portions of the shore are underwater only rarely and for short times when there is a storm surge.

Clearly, the tidal range is important. In areas with a very low tidal range, the vertical space covering the various flooding frequencies is small, which severely limits the spaces where plants with specific flooding tolerances can grow. Conversely, in areas with a relatively high tidal range, the space for the various micro-environments is larger, and it is easier to observe the zones of a plant community.

Areas with a very low tidal range present another set of problems. Oceanographers and coastal engineers can predict years in advance the basic astronomical tides that depend on the geometry of the movements of the sun, moon, and Earth. However, the observed water level often differs from the predicted tide, since other factors, especially wind and barometric pressure, play important roles. In places with low astronomical tides, the weather tides can dominate the system. If a storm surge is 1 foot (30 cm) and the mean tidal range is 10 feet (3 m), the difference is only 10 percent. If the surge happens at the time of low tide, the only difference will be a slightly longer time of flooding for part of the intertidal zone. However, if the mean tidal range is only 1 foot, not only is the weather tide equal to 100 percent of the tidal range, but the intertidal zone will be totally submerged. Also, in "microtidal" areas, the weather overprint on the water level can destroy the regularity of flooding and exposure. Some plants that can tolerate being underwater for a few hours and then exposed to the air for a few hours may not survive protracted flooding.

Figure 7.3 The bank of a marsh channel near low tide, with smooth cordgrass (*Spartina alterniflora*) growing on the top of the mud. The slight color change on the stems shows how high the water reaches at high tide. The shrubs and tree that live at elevations above the normal reach of the tide are in the background. (Photograph by the author)

In tidal marshes along the Atlantic and Gulf coasts of the United States, smooth cordgrass (*Spartina alterniflora*) has a high tolerance for submergence. It grows between roughly the mean tide level and just below the level of mean high tide, perhaps to the elevation of neap high tides (figure 7.3). The height of smooth cordgrass generally relates to the tidal range. While the plant's lower portions can tolerate being underwater for longer periods than the upper portions, some parts of the plant must remain above water except for occasionally being submerged during storms.

In a sequence from the edge of the ocean toward higher ground, the different plant species are progressively less tolerant of submergence. Moving higher into the marsh, the shorter form of smooth cordgrass becomes common. Then the vegetation transitions to saltmeadow cordgrass (*Spartina patens*), which is flooded only during spring tides and storms.

Black needlerush (*Juncus roemerianus*), from New Jersey south, and, mostly north of Delaware, saltmeadow rush (*Juncus gerardii*) populate the region slightly above saltmeadow cordgrass. *Juncus* occurs in a short vertical range that can be wide, provided the slope is very gentle. In New England, pickleweed or glasswort (*Salicornia virginica*) and saltgrass (*Distichlis spicata*) grow between the *Spartina* and *Juncus* marshes.

Because tidal marsh plants depend on the wet and dry balance of submergence and emergence, any change in sea level can alter the marsh. In many settings, marshes grow vertically by rising on top of themselves. As the plants grow and die, some of the detritus decomposes and piles on the marsh surface. The plants themselves hinder the flow of water, allowing some suspended sediment to settle. The elevation of the marsh surface rises slowly, on the order of 0.04 inch (1 mm) a year. To some degree, the compaction of the accumulated plant matter and inorganic sediment counterbalances the accumulation of new material. Over time, the increasing weight of decaying plant remains on the top of the marsh presses down on the underlying material, squeezing and compressing it. This compressed marsh material is called *peat*, and the natural compression of the marsh under its own weight sometimes is termed *autocompaction*. As the volume of the subsurface matter decreases, the marsh surface moves downward, perhaps compensating for the upward *accretion* of plant material and sediment. A changing sea level further complicates the situation. The combination of the land subsiding from the natural compaction of the marsh under its own weight, the accretion of the marsh surface, and changes in the eustatic (worldwide) sea level is the change in the local level of the sea relative to the land. The rate of change of the relative sea level at a salt marsh is important to the fate of the marsh.

If the marsh surface accretes more rapidly than the local sea level rises, eventually the marsh surface will be too high for the existing plant community, which needs regular submergence to survive. It will shift toward plants that prefer a drier environment. An area dominated by smooth cordgrass will change to a marsh of saltmeadow cordgrass. Observed from above over a long period, the marsh will appear to move toward the water. Plants that need less submergence will grow over plants that tolerate being underwater more often and for longer periods.

If the marsh surface accretes at the same pace that the local sea level rises, the plant species will continue to grow in the same places (laterally), but the entire marsh system will become thicker. Over thousands of years, marsh peats can grow tens of feet thick. If the local sea level rises more rapidly than the marsh surface accretes, the limits of tidal flooding shift inland. The plants of the lower marsh will appear to move toward the land and encroach on what had been the high marsh. But if the sea level rises faster than the marsh can grow or migrate, the marsh will not survive. As the plants die, the marsh likely will become open, shallow water or mudflat. The wholesale loss of tidal marsh that would accompany an increased rate of sea-level rise equates to the destruction of an important habitat with crucial environmental functions, such as filtering surface water and providing a flood buffer.

The sediments of the marsh are a complex environment. The great prevalence of silts and clays greatly limits the permeability of the marsh surface, so there is little active flow of water through the "body" of the marsh. Plant roots, decaying plant matter, and bacteria and other organisms consume the oxygen in the water that was trapped when the mud was deposited, leaving an oxygen-starved (anoxic), "reducing" environment. One of the characteristics of a reducing environment is the presence of hydrogen sulfide (H_2S), whose rotten-egg, marsh-gas smell assaults us as we trod through marshes and mudflats. If acidic water with a high level of reduced iron should flow through the marsh and encounter an oxygen-rich area, the iron will oxidize, forming the minerals goethite and limonite, which are often known as bog iron. Although deposits of bog iron are poor ores for commercial use, they have been mined as a source of iron.

We cannot end this chapter without mentioning mudflats. This unvegetated area occurs seaward of the lowest extent of smooth cordgrass. Although the pervasive odor of hydrogen sulfide and the slippery, sloppy nature of the flat are unappealing to most people, a rich variety of animals dwells in the mud. Clams and worms live within the anoxic area, but extend parts of their bodies into or maintain burrows that reach up to the water that flows freely when the flat is submerged by the tide. The surface layer of oxygenated sediment, usually only a fraction of an inch thick, is discernible as a light-colored zone on top of the dark, anoxic sediments. The chemical reactions, better described as bio-geo-chemical reactions, that occur in the oxygen-starved reducing zone are complex. As an example, microscopic particles of metal sulfides, such as iron sulfide (FeS_2), better known as pyrite or fool's gold, frequently form in mudflats and marsh muds. Recently, I saw a large clam with a golden shell; a thin patina of pyrite had formed on the surface of the shell, changing its normal chalky gray appearance to a dirty gold color.

* * *

On almost every trip to the beach, we encounter sand dunes, tidal salt marshes, and, perhaps, mudflats. Dunes develop where there is an abundance of sand and wind, and they can range from insignificantly small mounds of sand to immense deposits. Behind a beach, dunes are both a reservoir of sand with which nature replenishes an eroding beach and a physical buffer between a storm-tossed ocean and more shoreward regions. The plants that live in and around a field of dunes have to be specialized to tolerate the harsh, arid, sun-baked environment and only intermittent access to abundant water.

In comparison, the plants of a tidal salt marsh are specialized to tolerate salinity and varying periods of submergence. The local tidal range determines the vertical range of the different species. Although marshes grow vertically by incorporating the remnants of decaying plants, potentially requiring the vegetation to change over time, they also have to respond to changes in sea level and subsidence of the marsh surface. Depending on the combination and scales of the various processes, marshes can remain static, migrate landward or seaward, or fall apart, leaving an unvegetated, intertidal mudflat. Although a lot of water is contained between the grains of sediment in marshes and mudflats, the very low permeability of the fine-grained deposit greatly restricts flow. As a result, bio-geo-chemical reactions use up the oxygen, leaving a foul-smelling, anoxic environment.

8

SEA LEVEL AND SEA-LEVEL RISE

Why does sea level change? The concepts of sea level and sea-level change are complex and can be confusing. What does sea-level change, more usually called sea-level rise, mean? How is sea level measured at any one place? How is it measured throughout a region or around the world? How is the change in sea level measured, and why does it matter?

When you spend a day at the beach, without realizing it, you observe sea-level change on two time scales: the second-to-second scale of waves and the hour–to-hour scale of tides. These two processes highlight the problems of studying sea level. In one very real sense, sea level is the elevation of the water surface at any instant in time. So in the few seconds of a wave period, sea level can vary by several feet. Sea level fluctuates seasonally in response to changes in temperature. Barometric pressure, wind, and the 18.6-year tidal cycle also influence sea level. Broader changes in climate, whether natural or augmented by human activities, influence sea level on scales of several decades or centuries. Other oceanographic processes, such as El Niño–Southern Oscillation (ENSO) and the North Atlantic Oscillation (NAO), affect sea level as well. Another factor is regional or local land movement, which can be abrupt or gradual. The sudden shift of the land during an earthquake can, in seconds, make harbors shallower or deeper, changing the local sea level relative to the surrounding land by changing the position of the land in relation to the ocean. Slower processes gradually lift or lower the land relative to the sea. Seafloor spreading affects sea level over many millennia.

A term that bears on changing sea level is *eustatic sea level* (*eustacy*). The *Glossary of Geology* defines "eustacy" as "the worldwide sea-level regime and its fluctuations caused by absolute changes in the quantity of sea water." Through the past several million years, the growth and shrinkage of the polar ice caps have been the major mechanisms of eustacy. A slower but nonetheless real process is the addition of "new," primarily volcanic, water to the global hydrologic cycle and eventually to the oceans. Icy comets that hit Earth also add new water to the hydrologic system. Although technically not changes in the quantity of water in the oceans, changes in the volume and shape of the ocean basins influence the level of the sea surface. Sedimentation partially fills the ocean basins and, in so doing, displaces the water upward. Seafloor spreading and subduction alter the size and shape of the oceans. These processes usually are considered eustatic changes in sea level, since they have worldwide impact.

When people say "sea-level rise," what they mean is the rise through time in a tidal datum such as mean higher high water or mean tide level, both of which are determined by averaging readings over a 19-year period. According to the National Oceanic and Atmospheric Administration publication *Tidal Datums and Their Applications*, a 19-year period is used because it is the full-year interval closest to the 18.6-year astronomical tidal cycle. Tide gauges measure the water level inside a stilling tube to filter out, or "still," the rapid variations in water level caused by waves.

To determine regional sea level, coastal engineers survey a network of several tide gauges to ensure that they do not move up or down relative to one another and to tie them to a fixed reference point or datum. This fixed datum causes yet more confusion. It usually is a "geodetic" datum (although it may use the term "sea level"), which is closely related to the shape of Earth, the *geoid*. Finally, satellites are important tools in measuring both the elevation of the ocean's water surface and the shape of Earth.

When considering the consequences of sea-level rise, remember that the practical concern usually is with *relative sea level*, the level of the ocean as it meets the shore at any particular place, such as a town's harbor or a resident's waterfront property. Changes in relative sea level are the sum of the absolute, or eustatic, change; the regional or local change that results from the elevation or subsidence of the land surface due to other factors, such as faulting; and the several other processes that alter the water level. While trends in worldwide sea level are very important, they are difficult to determine.

At Juneau, Alaska, the historical record of sea level shows the impact of local tectonic uplift on relative sea level (figure 8.1). Although the eustatic sea level is rising, the relative sea level at Juneau is dropping. The local uplift of the land causes the apparent fall in sea level in this part of Alaska's Inland Passage. The importance of local tectonics on relative sea level is especially clear when the water-level records for Juneau are compared with those for Ketchikan (figure 8.2), which is only 230 miles (370 km) from Juneau. The 80-year trend at Ketchikan is for a nearly stable relative sea level, falling at only 0.06 foot per century (0.19 mm/year); at Juneau, it is falling at about 4.24 feet per century (12.9 mm/year). Also, recall that both of these rates are contrary to the global trend.

Tide gauges are relatively modern inventions. According to Bruce C. Douglas, although a few tidal records go back to the seventeenth century,

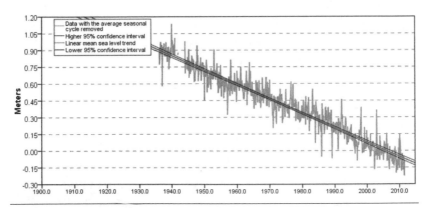

Figure 8.1 The record of sea level from 1936 through 2010 at Juneau, Alaska. The long-term trend depicts sea level as falling by 4.24 feet a century (12.92 mm/yr). (From National Oceanic and Atmospheric Administration, Sea Levels Online: Sea Level Trends, http://co-ops .nos.noaa.gov/sltrends/sltrends.shtml)

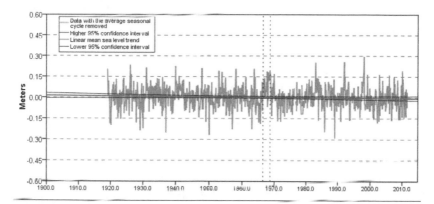

Figure 8.2 The record of sea level from 1919 through 2010 at Ketchikan, Alaska. The long-term trend depicts sea level as falling by 0.06 foot a century (0.19 mm/yr). The dashed vertical lines bound a period with questionable data. (From National Oceanic and Atmospheric Administration, Sea Levels Online: Sea Level Trends, http://co-ops.nos.noaa.gov/sltrends/sltrends.shtml)

the longest tide-gauge records began in Great Britain in the 1830s. The earliest records in the United States are at San Francisco, California, and The Battery in New York City, from the 1850s, although data gaps and other problems shorten the periods of continuous data. Most tide gauges have been in place for less than 60 years, barely three tidal epochs. Table 8.1 presents the sea-level trends for several locations in the United States, and figure 8.3 shows the tide-gauge data for the century-long record at Baltimore, Maryland.

* * *

Over the past several million years, ice has controlled major, worldwide changes in sea level. The contraction and expansion of Earth's ice sheets has caused sea level to rise and fall. As the climate cooled, the normal hydrologic cycle was interrupted, as snow did not melt and flow back into the ocean. Hence water shifted from the ocean to the land and sea level fell. Thousands of years later as the globe warmed, there was more melting than there was snowfall, and water that had been stockpiled in the ice

TABLE 8.1

Recent Sea-Level Trends

Site	State	Years of record	mm/yr	ft/ century
Eastport	Maine	78	2.0	0.66
Bar Harbor	Maine	60	2.04	0.67
Portland	Maine	95	1.82	0.60
Boston	Massachusetts	86	2.63	0.86
Woods Hole	Massachusetts	75	2.61	0.86
Nantucket Island	Massachusetts	42	2.95	0.97
Providence	Rhode Island	69	1.95	0.64
New London	Connecticut	69	2.25	0.74
Bridgeport	Connecticut	43	2.56	0.84
Montauk	New York	60	2.78	0.91
Port Jefferson	New York	36	2.44	0.80
New York City (The Battery)	New York	151	2.77	0.91
Sandy Hook	New Jersey	75	3.90	1.28
Atlantic City	New Jersey	96	3.99	1.31
Lewes	Delaware	88	3.20	1.05
Baltimore	Maryland	105	3.08	1.01
Kiptopeke	Virginia	56	3.48	1.14
Sewells Point	Virginia	80	4.44	1.46
Beaufort	North Carolina	54	2.57	0.84
Wilmington	North Carolina	72	2.07	0.68
Myrtle Beach (Springmaid Pier)	South Carolina	50	4.09	1.34
Charleston	South Carolina	86	3.15	1.03

Site	State	Years of record	mm/yr	ft/ century
Fort Pulaski	Georgia	72	2.98	0.98
Fernandina Beach	Florida	110	2.02	0.66
Miami Beach	Florida	51	2.39	0.78
Key West	Florida	94	2.24	0.74
St. Petersburg	Florida	60	2.36	0.78
Apalachicola	Florida	40	1.38	0.45
Dauphine Island	Alabama	41	2.98	0.98
Grand Isle	Louisiana	60	9.24	3.03
Sabine Pass	Texas	49	5.66	1.86
Galveston (Pier 21)	Texas	99	6.39	2.10
Port Isabel	Texas	63	3.36	1.19
San Diego	California	101	2.06	0.68
Los Angeles	California	84	0.83	0.27
San Francisco	California	110	2.01	0.66
Alameda	California	68	0.82	0.27
Crescent City	California	74	−0.65	−0.21
Astoria	Oregon	82	−0.31	−0.10
Neah Bay	Washington	73	−1.63	−0.53
Seattle	Washington	109	2.06	0.68
Honolulu	Hawaii	102	1.50	0.50

Source: Data from http://co-ops.nos.noaa.gov/sltrends/msltrendstable.htm and http://co-ops.nos.noaa.gov/sltrends/msltrendstablefc.htm (accessed June 12, 2011).

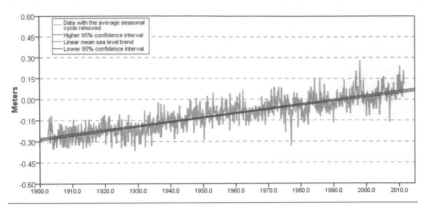

Figure 8.3 The record of sea level from 1902 through 2010 at Baltimore, Maryland. The long-term trend depicts sea level as rising by 1.01 feet a century (3.08 mm/yr). (From National Oceanic and Atmospheric Administration, Sea Levels Online: Sea Level Trends, http://co-ops.nos.noaa.gov/sltrends/sltrends.shtml)

caps returned to the ocean. The glacio-eustatic movements of sea level typically range up to about 460 feet (140 m) and occur over many thousands of years. Scientists have calculated the differences in the volumes of ice and seawater during glacial episodes and at various levels of the sea.

A parallel phenomenon, with a quicker rate of response and a smaller vertical scale, is the change in water volume when the ocean's waters expand or contract as the temperature changes. The density of water is greatest at about 39°F (4°C). As the water temperature moves away from that value, the volume of water increases; conversely, as the temperature moves toward 39°F, the volume decreases. Virtually all of that volume change displaces the upper surface of the ocean, since the bottom and "sides" are fixed. The temperature-controlled change in sea level between glacial and interglacial epochs is a few feet. The warming of the surface water of the ocean during the past two or three decades is responsible for a portion of the recent rise in sea level. On a smaller scale, temperature changes cause seasonal variations in sea level. The yearly cycle of sea level at Baltimore follows the seasons, with the lowest level occurring during the coldest months (figure 8.4). Other seasonal differences, such as in barometric pressure and wind direction, also contribute to the annual cycle.

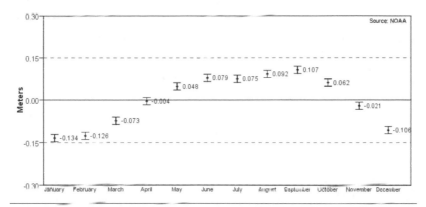

Figure 8.4 The average seasonal mean-sea-level cycle with 95 percent confidence intervals at Baltimore, Maryland. The water level is highest in the late summer, August and September, and lowest in the deep winter, January and February. (Data from National Oceanic and Atmospheric Administration, Sea Levels Online: Sea Level Trends, http://co-ops.nos.noaa .gov/sltrends/sltrends.shtml)

Further complicating sea level, Earth's crust sits atop a viscous mantle that flows slowly as the pressure above it changes. When a substantial thickness of ice accumulates, that regional increase in load presses the crust deeper into the mantle. Conversely, when the ice melts and the load is reduced, the crust rebounds upward. This phenomenon is called *isostasy*. As the ice sheets on land grow or melt, the load transfers from and to the ocean basins similarly affect the elevation of the seafloor. The subsidence or uplift is not instantaneous, but lags substantially behind the change in load.

The local history of relative sea level can be complex. That along the coast of Maine is a good example. About 14,000 years ago, when the eustatic sea level was more than 300 feet (100 m) lower than it is today, the weight of the ice was sufficient to warp Earth's crust down so far that areas then at sea level are today 230 feet (70 m) above sea level. The region between today's 230-foot elevation and the shore would have been flooded, but the sea abutted the ice sheet and not the land. Then as the ice retreated inland, the coastal area was inundated. The uplift that accompanied the removal of the ice load eventually raised the shallow areas above the present-day sea level. But the rapid, eustatic sea-level rise again inundated parts of the coast. Since the ice began to retreat, parts of coastal Maine have been under ice, under the sea, above sea level, and below sea level. The "postglacial rebound" of regions that were depressed under the ice continues. According to the Maine Geological Survey, a series of small earthquakes in April and May 2011 may have been a result of adjustments to the retreat of the ice cap long ago.

The geographic extent of glacially driven subsidence and rebound extends beyond the regions that were covered with glaciers. Tens to hundreds of miles from the ice, a bulge formed as the crust moved under the ice cap and gooey, viscous mantle flowed outward. The bulge began to relax as the ice cap retreated. The region that was lifted on the bulge began to subside. This relaxation continues, thousands of years after the ice melted, and likely contributes to the high rate of relative sea-level rise around the mouth of Chesapeake Bay.

Scientists have tested the concept that a load on Earth's surface can depress the crust. They compared precise topographic surveys conducted around Hoover Dam and Lake Mead, on the Nevada–Arizona border, between 1936, when the dam was completed, and 1950. The work indicated that an area about 37 miles (60 km) in diameter centered on the deepest part of the reservoir sank at least 2.7 inches (70 mm). The shorelines of ancient glacial lakes also demonstrate the consequences of glacial rebound. At the end of the last ice age, Lake Hitchcock, 200 miles (300 km) long, filled the Connecticut River Valley from Rocky Hill, Connecticut, to central Vermont. Today, the shoreline of former Lake Hitchcock, which originally was level, tilts up to the north at about 4 feet per mile (0.75 m/km), showing that the northern areas had sunk farther than the southern areas and since the retreat of the ice sheet have risen to their present level.

Just as changes in the distribution of ice sheets alter the distribution of load on Earth's crust, changes in the water depth in the ocean basins alter the load on the sea. If 1 cubic foot (0.03 m³) of seawater weighs about 64 pounds (29 kg), a 450-foot (137-m) rise in sea level would change the load at the seafloor by about 28,800 pounds per square foot (200 lb/in² [14 kg/cm²]). This would slightly depress the crust under the sea. The mass of the ice in the continental ice sheets exerts a gravitational pull on the ocean waters. Because the attraction of gravity decreases with the square of the distance, the primary effect on sea level is close to the ice sheets and moves as the ice sheets shrink or expand. Several other factors stem from the redistribution of the mass of water from the oceans to the ice sheets and the resultant changes in load on the crust that have consequences far away from the ice.

* * *

The complexity of long-term processes that change the sea level then force a question: How do we determine the sea level at some time in the distant past? We need at least three pieces of data:

- Something that indicates the location of an ancient sea level
- The age of that artifact
- The present elevation of the relic

Once we find the first item, the second and third are relatively easy to obtain.

The usual method of estimating an age is radiometric dating. Radiocarbon, or carbon-14 (^{14}C), dating is a common technique for aging animals and plants that lived within the 20,000 or so years since the last glacial maximum and the following rise of sea level, although the technique works back to about 45,000 years. Other radiometric-dating methods are also used. Uranium-series dating, for example, is used to determine the age of corals and has a much longer useful period, about 500,000 years, enabling the calculation of the ages of older high stands of sea level.

A chart or map can show elevation, but a survey is more accurate. The real problem is finding a relic that indicates sea level. Accurate and precise markers of sea level are uncommon. For the sake of discussion, let's use mean high water (MHW) as a reference. The beach crest, where the angle breaks between the relatively flat berm and the seaward-sloping beach face, occurs slightly above MHW. That's fine for today, but it is very difficult to locate ancient submerged beach crests. In areas where the relative sea level has fallen, "fossil" beaches sometimes are stranded as the sea moves away from them, but this circumstance is rare. Even where fossil beaches are apparent, how can we determine their age? If we plot sea level versus time since the last glacial maximum, radio-carbon analysis usually can show the age of organic matter. But the beach is notoriously barren of plants and in-dwelling animals. Thus old beaches are not particularly good tools for assessing past sea level.

Mobile animals—fish, for example—even though they provide ample material for determining age, do not work well. Finding a fish skeleton tells us almost nothing about sea level other than that the area where the skeleton was found probably was below MHW, unless the dead fish was carried to a higher place by a scavenger or deposited there during a storm surge or another process not tied to sea level. To be useful, a fossil has to be discovered where it lived and has to be an animal or a plant that provides precise information about sea level. Some species of clams and oysters might be helpful; after all, they can live in intertidal areas. But they also can live in deeper water, as much as 30 feet (10 m) and more, so their utility as a precise indicator of sea level is limited.

Marsh vegetation offers another line of evidence. Several plants live in zones that are closely related to tide levels. For example, smooth cordgrass (*Spartina alterniflora*) is restricted to the region between about midtide and spring high tide. Saltmeadow cordgrass (*Spartina patens*) occurs just above smooth cordgrass. Other plants grow in specific habitats that are defined by the frequency and duration of tidal flooding. Various plants provide information about sea level while they are alive and thus are easily recognizable. Unfortunately, after they die and only some root materials remain, identification is difficult or impossible. Another problem is that the aging root and plant matter, the peat, can compress under its own weight. In an area with a sustained rise in sea level, the marsh peat often grows very thick. The weight of the overlying material squeezes the older plant matter. In this process of autocompaction, everything above the base of compression moves downward from the level of deposition. Also, plant roots grow into older sediments. Thus an age date on the root matter really yields a date slightly younger than the age of the sediments into which the roots grew.

To minimize these confounding factors, the usual procedure is to sample the very bottom of the marsh peat where it sits atop a firm, older sediment layer. The top of the older stratum is the surface that the rising sea crossed. Thus the age of organic matter immediately on top of an older surface would yield the minimum date when the rising sea would have inundated that level. By taking a series of cores in a line up the slope and dating peat taken from where it touches the older layer, the ages depict the change in sea level through time. Scientists expect that technical reports will show "error bars" on the curve's data points. Error exists in two dimensions: time, usually shown as the uncertainty around the radiometric dating technique, and elevation.

Recently, researchers have begun to impose even more controls on the raw data by plotting the lowest occurrence of land material as well as the highest occurrence of marine indicators. Thus sea level is confined to a bracket between the terrestrial and marine samples.

Since radiocarbon dating is limited to organic matter, it would be helpful to have means to determine the age of other materials. Luminescence dating is another way to measure the age of some inorganic, sedimentary matter, especially quartz and feldspar, minerals that are common beach components. The analysis determines the time since the minerals were last exposed to sunlight or last heated to about 900°F (500°C). While heating works well for archaeologists in dating ceramics, the time since a sample was exposed to sunlight is useful to geologists because it indicates when other sediments were deposited on top of the sample.

Some less direct methods can estimate past sea levels, and a few techniques can provide data far back into the Pleistocene epoch (2.5 million–11,700 years B.P.) and beyond. By using a proxy to estimate changes in ocean volume, it is possible to estimate past sea levels. The ratio of two isotopes of oxygen, oxygen-16 (^{16}O) and oxygen-18 (^{18}O), is a widely used proxy for sea levels extending back 250,000 years or more. In seawater, ^{16}O occurs about 500 times as often as ^{18}O. Water molecules with ^{16}O evaporate more easily than the heavier ^{18}O-bearing molecules ($H_2^{16}O$ has a lower atomic weight than $H_2^{18}O$). Thus the ice caps, which grow because the quantity of snow exceeds the rate of melting and runoff, are disproportionally rich in ^{16}O, leaving the oceans similarly enriched in ^{18}O. This ratio offers two, complementary routes of investigation. But it indicates only past worldwide, or eustatic, sea level and provides no details of local relative sea levels. Oxygen isotopes are very widely used in studies of past sea levels and climates, but, as we should expect, it is not always as simple as just determining the ratio of ^{18}O to ^{16}O. Other data, such as estimates of the temperature of the bottom water, contribute to the calculations.

The Arctic and the Antarctic are windows into Earth's history. Scientists count the greatly compressed, annual layers of snow in an ice core to ascertain the time since any particular stratum was deposited. They also can analyze a sample of the ice from a dated layer to determine its ratio of ^{18}O to ^{16}O. By repeating this procedure for many layers within a core, the scientists can construct a plot of the isotope ratio versus time. This plot mimics a plot of sea-level change, as a higher proportion of ^{16}O indicates a lower sea level. The complementary procedure is to extract shells from strata of sediment in cores from the ocean floor. The shells can come from the microscopically small animals known as foraminfera as well as from larger animals. Most shells are calcium carbonate ($CaCO_3$), made from elements removed from seawater. The shells' isotopic makeup represents the conditions in the ocean when the animals died. (The age of the sediment layer where the shells were found can be determined independently, perhaps from dating the shell itself.) Shells with a relatively higher level of ^{18}O than the shells above or below them indicate that the animals lived during a glacial (colder water) period. This is because much of the ^{16}O that was preferentially evaporated from the sea was deposited in the ice caps, depleting the sea in ^{16}O and enriching it in ^{18}O. The many studies of the oxygen-isotope ratios in both ice and sediment cores from around the world show a recognizable sequence of variation that has been divided into a series of "oxygen-isotope stages." The series extends hundreds of thousands of years back into geologic time.

Our understanding of the change in sea level since the peak of the last glaciation has become increasingly detailed. Not that many years ago, geologists interpreted the postglacial rise in sea level as a smooth curve. The curve rose rapidly from the minimum sea level about 15,000 years ago until about 6,000 years ago, when the rate of rise slowed (figure 8.5).

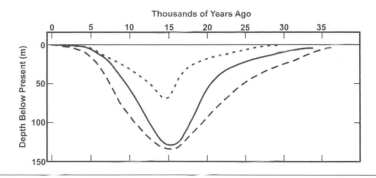

Figure 8.5 The history of sea level along the Atlantic coast of the United States during the past 35,000 years. The dashed lines indicate the range of estimates, and the solid line between them represents a likely sea-level curve. (Modified from J. D. Milliman and K. O. Emery, Sea levels during the past 35,000 years, *Science* 162 [1968]: 1121–1123)

Scientists have refined the shape of the curve and debated whether sea level has oscillated within a few feet of the present level during the past 5,000 years. More recent studies show that sea-level rise has not been smooth but has occurred in a stair-step manner: short intervals of rapid rise, sometimes extremely rapid, separated by longer periods when sea level varied slightly either up or down. Figure 8.6 presents the history of sea-level change in the Yellow Sea, between eastern China and Korea. Figure 8.7 is a record of sea level for a much longer period.

* * *

Turning from the past to the future, we want to know how high and how fast sea level will rise not only in our lifetimes, but also in our children's and grandchildren's. Although we can chart the history of sea level with a high degree of accuracy, predicting future sea levels is less sure. There is wide, virtually universal, agreement among scientists that sea level is rising and, most likely, will continue to do so for the foreseeable future and probably at a faster rate. There is some question that we may be approaching the switching point where Earth's climate will begin the change toward the next glacial episode; however, there is genuine concern that humans may have modified Earth's climate so much that the

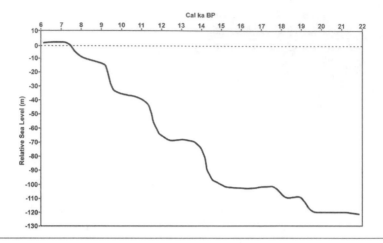

Figure 8.6 The stepped sea-level curve for the Yellow Sea, from 22,000 to 6,000 years before the present. (Modified from J. P. Liu, Post-glacial sedimentation in a river-dominated epicontinental shelf: The Yellow Sea example [Ph.D. diss., Virginia Institute of Marine Science, College of William & Mary, 2001], and J. P. Liu, J. D. Milliman, S. Gao, and P. Cheng, Holocene development of the Yellow River's subaqueous delta, North Yellow Sea, *Marine Geology* 209 [2004]: 45–67)

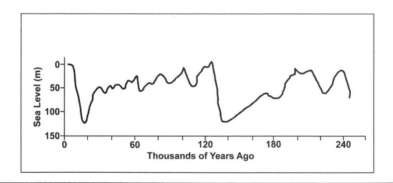

Figure 8.7 The worldwide sea-level curve for the past 250,000 years. (Modified from J. Chappell and N. J. Shackleton, Oxygen isotopes and sea level, *Nature* 324 [1986]: 137–140)

geologic mechanisms may no longer function and Earth will continue to warm.

The issue of human influence on climate is based, in large part, on the concept of "greenhouse gases," as described in both the popular and scientific media. Several studies have documented increasing levels of several gases, such as carbon dioxide (CO_2), in Earth's atmosphere. The majority of expert opinion is that the increase in the gases is causing Earth to warm, and that much of the increase is due to anthropogenic (human-caused) activities, primarily burning fossil fuels. While most scientists agree on the concept of global climate change, there is less certainty about the rates at which Earth will continue to warm and sea level will rise. A small minority of scientists does not accept the concept of anthropogenic warming.

Projections of future changes in sea level range widely, from a low of approximately the rate of the past tidal epoch to highs of double or triple or more the present rate. In *Climate Change 2007*, the Intergovernmental Panel on Climate Change (IPCC) estimated that by 2100 the average global sea level is likely to have risen at least 7 inches (17 cm) and as much as 2 feet (60 cm). More recent studies indicate that sea level is likely to rise at least 3 feet (1 m) and, quite possibly, 5.25 feet (1.6 m) by the end of the twenty-first century. Beyond 2100, the warming of both the atmosphere and the ocean may have progressed to a level that would accelerate the melting of glaciers and ice sheets to extreme levels.

Why do we care about future sea level? A rising sea level presents physical and societal consequences. Although a rise of a foot (30 cm) or so within a generation might seem insignificant, it would cause real problems. A rise of 3 feet (1 m) or more by the end of this century would lead to significant changes to the coastal areas, resulting in major impacts on society. The lower Mississippi River Delta is at great risk. A small rise of sea level would inundate large portions of the rural areas around Chesapeake Bay in Virginia and Albermarle and Pamlico sounds in North Carolina. In these low-lying regions, people rely on shallow wells for drinking water and on septic systems for domestic waste. Increases in sea level affect both of these necessities. As sea level rises, the brackish water lens in the surface aquifer rises as well. More important, coupled with its higher level, the brackish water reaches farther inland where it can sour the well water that families use for drinking and cooking. The simple rise of the level of the groundwater can ruin septic drain fields. There is no easy or inexpensive fix for these problems. Constructing sewer or water systems is very costly, and in rural areas with a low population density, there are few families and businesses to absorb the cost.

If the rate of sea-level increase is sufficiently rapid, the growth of marshes may slow, resulting in the loss of wetlands and critical habitat. The great destruction of marsh habitat endangers animals that rely on tidal wetlands for any portion of their life cycles. The disappearance of wetlands as a consequence of rising sea level also increases the area of open water. Even though it may be shallow, the open water on the mud flats yields an increased fetch, thus potentially allowing larger waves that, in turn, would accelerate erosion.

Simple, though gradual, flooding of low-lying coastal areas is an inevitable consequence of sea-level rise. Less obvious, however, are storm flooding and storm surge. Coastal structures are built above a specific flood-frequency level. For example, the "50-year flood" is the water level with a 2 percent chance of occurring in any one year. But if sea level is rising at 1 foot (30 cm) a century, in 25 years, sea level will have risen 3 inches (8 cm) and have pushed the 50-year-flood level that much higher. Three inches might not seem like much, but in terms of the less frequent flood levels, it can make a huge difference, perhaps doubling the chance that the coastal building will be flooded.

Flooding is more complex than it may appear. As sea level rises, not only is land submerged, but the shoreline moves landward; geologists call this *transgression*. The physical aspects of erosion move onshore, with a loss of land and an enhanced potential for flooding. Economically, the disappearance of land can lower local tax rolls and alter permissible land use. If sea level rises gradually, its impacts might escape notice. However, storms remind us of the shifting base level. Spits and barrier islands become narrower because the shoreline moves inward from both the seaward and landward sides.

In late August 2005, Hurricane Katrina spotlighted the vulnerability of New Orleans, which depends on pumps to remove rain- and flood water from sections of the city that are below both river and sea level. For a combination of reasons, southern Louisiana is subsiding very rapidly. Relative sea level at Grand Isle, about 60 miles (100 km) south of New Orleans, is rising at about 3 feet (1 m) per century (0.4 inch [0.92 cm] per year). Thus, on average, a levee or flood wall designed to a specific elevation today will be about 4 inches (10 cm) too low in 10 years, about the time construction finishes.

In September 2003, Hurricane Isabel struck the Outer Banks of North Carolina and Chesapeake Bay. Although Isabel was only a minimal Category 1 hurricane when it hit Virginia—and it made landfall during neap tides—the storm surge reached or exceeded levels that had not been seen in 70 years. In southeastern Virginia, the storm surge flooded a major roadway tunnel when protective measures failed. That a relatively minor hurricane could have such major impacts—the tunnel was closed for weeks—should alert citizens and governments to changing conditions. In the 40 to 50 years since engineers designed the region's tunnels, local sea level has risen about 8 inches (22 cm). Adding storm tides to the higher general sea level increases the likelihood of flooding. The media have reported that subway tunnels in New York City could flood and Lower Manhattan could be submerged.

Around the world, the predictable consequences of sea-level rise are even more daunting than in the United States. Several low-lying island nations in the Pacific and Indian oceans will be drowned, displacing their populations. Many people in Bangladesh live very close to sea level. Even before the sea permanently inundates some areas, the increased reach of storm floods will endanger hundreds of thousands of people and force them to higher areas. This new class of needy refugees will tax national and global resources. Simple prudence suggests that nations, regions, and cities must plan to manage the potentially devastating consequences of sea-level rise and, where physically possible, should act to minimize the factors that exacerbate this very real problem. The cost of mitigation is substantial, but the potential costs of losses are enormous.

Many scientists believe that increases in the very near future will be dramatic. The stair-step history of sea-level rise illustrates past jumps that would be catastrophic should they occur now (see figure 8.6). The steeply stepped pattern began about 19,000 years ago, at the end of the last glacial maximum, and continued until about 7,000 years ago. The sudden, rapid rises of sea level most likely occurred as major ice sheets collapsed or as the natural ice dams that impounded major glacial lakes failed. These events released immense quantities of meltwater into the world ocean. The Melt Water Pulses caused sea level to rise so rapidly that some of the scientific literature refers to Catastrophic Rise Events. As an example, during the Melt Water Pulse between approximately 9,300 and 9,000 years ago, sea level rose about 50 feet (15 m), or about 1 foot (30 cm) every 6 years (1 m every 20 years).

The speed of sea-level rise during the Melt Water Pulses is incomprehensible in light of the present-day discussion of sea-level rise, which predicts an increase from 7 inches (17 cm) to 5.25 feet (1.6 m) by the end of the twenty-first century. The speed of sea-level rise between 9,300 and 9,000 years ago is more than three times faster than the highest estimate for the future.

The Catastrophic Rise Events provide an edge to concerns about contemporary sea-level rise. While there is ongoing debate about the rates and consequences of sea-level change over the next century or more, very few people consider the possibility of a Catastrophic Rise Event. The collapse of either or both of the Greenland and West Antarctic ice sheets would have drastic consequences. Some scientists consider these events to be distinct possibilities. Recent estimates indicate that the maximum rise in sea level from the total melting of the West Antarctic Ice Sheet would be slightly over 26 feet (8 m) and the thawing of the Greenland Ice Sheet would yield more than 21 feet (6.5 m). A Catastrophic Rise Event would have profound consequences extending far inland of the beach and immediate coast.

* * *

Sea level is an important concept, and sea-level change—especially modern sea-level rise—is a critically important phenomenon. In order to calculate sea-level change, we must be able to determine the level of the sea at different times. In doing so, we have to be aware that local subsidence or uplift of Earth's crust adds to changes in the worldwide (eustatic) sea level. Sea level varies on a range of time scales. Other than the daily fluctuations due to the tide, the most important time frame is that of decades and centuries associated with the cycles of glaciation and deglaciation. The most dynamic factor in varying the level of the world ocean is the large increase or decrease in the volume of ice on land, primarily in Antarctica and Greenland. Changes in the temperature of the ocean, which often are related to changes in the mass of ice, also contribute to sea-level rise or fall through expansion or contraction of the water.

There is no doubt that sea level is rising, but there are questions about the future rate of rise and some discussion of the reasons for the rise. A very strong preponderance of the scientific community accepts that human activities, mostly related to the burning of fossil fuels, play the major role in contemporary climate change, which is the major contributor to sea-level rise. The evidence indicates that the rate of rise will accelerate during this century. The societal consequences of sea-level rise, perhaps as much as 5.25 feet (1.6 m) by the year 2100, would be profound and would include displacement of the millions of people who live in low-lying areas around the world.

9
STORMS AND STORM SURGE

Storm surge is the difference between the predicted, astronomical tide and the actual water level (figure 9.1). Although we think that storm surges cause unusually high tides, they also can lower anticipated water levels. The observed water level, occasionally called a storm tide or weather tide, combines the storm surge and the predicted, normal water level.

Wind, waves, and barometric pressure create the storm surge. Sustained, strong winds blowing toward the coast push seawater, so it piles up on the shore. Additionally, Ekman transport causes the water immediately below the wind-driven surface to flow to the right of the wind (in the Northern Hemisphere). This can be important along the Middle Atlantic coast, since the strong winds of nor'easters (northeasters) transport water landward. Large waves in shallow water move water with them, causing a wave setup of the water surface. Water tends to bulge up under areas of low barometric pressure. In very rough terms, a decrease of 1 inch (of mercury) (34 millibars) of barometric pressure yields about a 1-foot (30-cm) rise of the water surface. The process, termed the *inverted barometer effect*, is crudely analogous to sucking water through a straw, with the water rising in the straw as the air pressure in the straw falls compared with that in the surrounding atmosphere. Strong offshore winds and high barometric pressure can cause unusually low water levels. Finally, as the volume of water moves into shallow water or narrowing estuaries, the

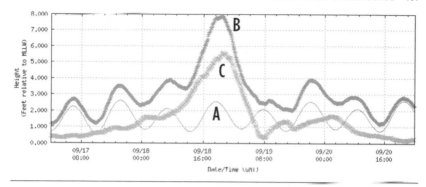

Figure 9.1 A plot of the (A) predicted and (B) observed water levels for Sewells Point (Norfolk), Virginia, for September 17–19, 2003, during Hurricane Isabel. The difference between the elevations (C) is the storm surge, which has a maximum of about 5.5 feet (1.6 m). (From National Oceanic and Atmospheric Administration, http://co-ops.nos.noaa.gov/data_res.html)

constriction forces the storm surge to rise, much as it does with the astronomical tide.

The high water levels of storm surges cause flooding and can allow the destructive effects of storm waves to reach higher elevations and farther inland. At the other extreme, unusually low water levels can expose greater areas of sand flats and have adverse consequences for shallow-water navigation, since depths are less than those depicted on nautical charts.

Because the storm surge is added to (or subtracted from) the normal water level, the impact often is related to the stage of the astronomical tide and to the tidal range. An elevated storm surge at the time of a spring high tide can be very destructive, while an elevated surge coincident with a normal low tide may have little consequence, especially if the magnitude of the surge is less than the tidal range. Similarly, a negative surge has more impact at low tide than at high tide. It is important to remember that it is the final elevation of the water surface that is important, not just the size of the surge.

Weather forecasts often estimate the height of the storm tide above mean lower low water (MLLW). While this is the correct frame of reference because the formal tidal datum usually is MLLW, the general public might be better served if a storm's water level were referenced to mean high water (MHW) or mean higher high water (MHHW). Most people living or working along the shore have a sense of where MHW is and can more easily visualize the increment above high tide than the larger increment above low tide. Also, we should remember that the forecasts of storm water levels are for the "still" water level. During a storm, waves are atop the surge and thus further increase its potential damage.

The highest storm surge ever recorded, 43 feet (13 m), was at Bathurst Bay, Australia, during a hurricane in 1899. According to the National Oceanographic and Atmospheric Administration, Hurricane Katrina in late August 2005 produced a record surge for the United States—at least 30 feet (9 m) along the Mississippi coast of the Gulf of Mexico. The previous American record was set in August 1969 by Hurricane Camille, which produced a storm surge in excess of 25 feet (7.6 m), also along the Mississippi coast. Storm surges cause most of the deaths associated with hurricanes and typhoons. In the very low coastal regions of Bangladesh and eastern India on the Bay of Bengal, tens or hundreds of thousands die from typhoons.

The storms to watch along the Atlantic and Gulf coasts of the United States are hurricanes and nor'easters. Hurricanes are "tropical" storms that form in tropical or semitropical areas; are centered on warm, humid air; and increase in intensity, with the growth of massive cumulus clouds and thunderstorms. Hurricanes usually strike from June through November, but on rare instances can occur in other months. Hurricane Zeta formed in December 2005 and dissipated in January 2006. Because of their power, hurricanes can be agents of massive coastal change. Usually, a hurricane's region of high winds is relatively compact. Since they tend to move rapidly, hurricanes seldom affect an area for more than one high tide, but the damage they inflict can be severe. As discussed in chapter 2, hurricanes are rated 1 to 5 on the Saffir-Simpson Hurricane Scale (appendix 3), depending on the strength of their winds.

The history of hurricanes reaching the Atlantic and Gulf coasts is very well documented. Eric Larson's book *Isaac's Storm* chronicles the hurricane that struck Galveston, Texas, in September 1900. Approximately

6,000 people lost their lives, mostly by drowning; many still consider that storm to have been the greatest natural disaster, at least in terms of deaths, to hit the United States. In *A Wind to Shake the World*, Everett Allen recounts the story of the Great Hurricane of 1938, which blasted Long Island and southern New England. It remains the "great storm" by which others that make landfall along the coasts of the Middle Atlantic and New England states are judged. In late August 2005, Hurricane Katrina devastated substantial portions of the Gulf Coast. Its record high surge brought brutal storm waves to Waveland and Bay St. Louis, Mississippi, as well as other coastal towns in Louisiana, Mississippi, and Alabama, causing massive destruction. The storm also led to the failure of the levee system that surrounds New Orleans, flooding vast areas of the city, destroying homes, and leaving many, many thousands of residents homeless.

Hurricane Irene assaulted the East Coast from North Carolina to Canada from August 27 to 29, 2011. Its path—it made landfall near Cape Lookout, North Carolina; moved back into the Atlantic Ocean near the North Carolina–Virginia border; made landfall, again, near Atlantic City, New Jersey; passed close to New York City; and finally moved inland—had the potential to make Irene an exceptionally destructive coastal storm. It was the first hurricane to threaten the New York metropolitan area since Gloria in 1985, and a storm surge could have inundated densely populated, low-lying areas of the city and its suburbs and flooded the subway system. While Irene did cut the barrier islands of North Carolina's Outer Banks in several places, destroying the road and isolating several communities, it lost strength more quickly than predicted, sparing northern New Jersey and the New York City area much of the anticipated damage, but causing much devastation in upstate New York and inland New England.

Nor'easters are extratropical storms that usually develop in the late autumn, winter, and spring. The term "northeaster" comes from the wind direction. As wind circulates counter-clockwise around a low-pressure area, the winds from a low moving north along the coast first approach shore from the northeast. Nor'easters spin up along the Middle Atlantic coast of the United States from interactions of frontal activity, the jet stream, and areas of low atmospheric pressure. Although seldom reaching hurricane intensity, they frequently have winds of 30 to 50 knots

(35–58 mph [15–25 m/sec]) across large areas, and they sometimes move slowly. Thus a major nor'easter can assault a coastal region for a full day or more. This means that the storm surge and high waves can attack the shore through multiple high tides, which substantially increases the damage to beaches. The damaging storms of November 2009, October 1991, and March 1962 each attacked the shore through three or more high tides.

Because nor'easters are often slow moving and cover large areas, they can cause a greater dollar value of damage than hurricanes. Even though a nor'easter's wind speed may be only half, or less, than that of a hurricane, the prolonged assault along a large stretch of coastline can be devastating. The Ash Wednesday Storm of early March 1962 ravaged much of the Middle Atlantic coast. New Jersey suffered the attack on March 6 to 8, with offshore waves over 30 feet (10 m) high. The storm occurred during spring tides, bringing with it a surge of about 3 feet (1 m) that lasted through four high tides. The Halloween Nor'easter or "Perfect Storm" of October 1991 is another example of a significant nor'easter that attacked the Middle Atlantic and New England coastal region with hurricane-like intensity. In November 2009, the remnants of Hurricane Ida merged with a growing nor'easter to form Nor'Ida, a storm that hit the Middle Atlantic coast with strong winds and a storm surge that persisted through eight high tides. At the peak of the storm, water levels approached those of Hurricane Isabel in 2003.

The storm track, or the path a storm takes, affects its impact on the coast. Whether a nor'easter or a hurricane, a storm that generally parallels the shore may have more total impact than one that approaches straight on (figure 9.2), but the small area where the storm crosses the shoreline will suffer the most intense damage.

The frequency with which storms attack also influences the damage to a beach. If a second storm hits the same section of coast within a week or so of an earlier storm, the beach will not have recovered. If the first storm degraded the natural buffers, even if the second storm is not as intense as the earlier one, it may be more damaging. Thus the four factors that control the impact of a storm are

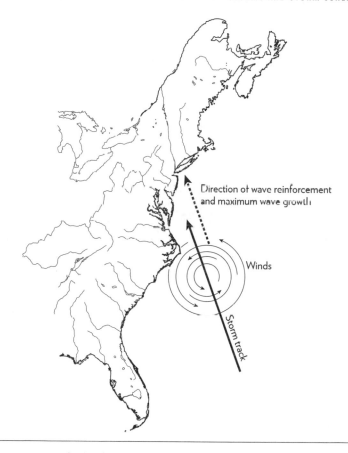

Figure 9.2 A storm, whether hurricane or nor'easter, moving generally north along the East Coast of the United States potentially causes more damage to Long Island and the southern shore of New England than to the coast of the Middle Atlantic region. (Illustration redrawn by Network Graphics from an original by the author)

- The intensity of the storm
- The storm track relative to the coast
- The duration of the storm, or the speed of its passage
- The storm frequency, or the time since the previous onslaught

Hurricane Katrina shows the importance of the storm track (figure 9.3). While the news media concentrated on the catastrophe that befell New

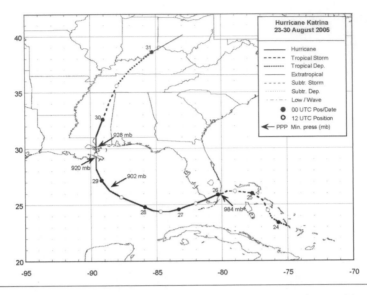

Figure 9.3 The storm track of Hurricane Katrina, August 23–30, 2005. The eye of the storm made landfall near the Louisiana–Mississippi border while moving due north. New Orleans, just to the west of the eye, escaped significant damage from the immediate storm winds and surge, while the coast of Mississippi, just to the east of the eye, suffered extreme damage from storm winds and surge. The greatest storm surge, about 20 feet (6 m), occurred near the Mississippi–Alabama border. (From R. D. Knabb, J. R. Rhome, and D. P Brown, Tropical cyclone report, Hurricane Katrina, 23–30 August 2005 [December 20, 2005], National Weather Service, National Hurricane Center, http://www.nhc.noaa.gov/pdf/TCR-AL122005_Katrina.pdf)

Orleans, most of the damage to that city came from the post-storm failure of the levee system. The sections of the city below sea level flooded when water poured through the broken levees. Katrina was moving northward when it made landfall slightly east of New Orleans; thus the city and much of the delta area did not suffer the highest winds, waves, and storm surge. East of the track line, however, high winds and large waves borne on the storm surge devastated the coasts of Mississippi and Alabama. Communities such as Waveland and Bay St. Louis were obliterated because they were both flooded by the surge and pounded by the great waves.

* * *

As we think about storms, we have to consider the potential consequences of climate change and sea-level rise. Some climatologists have suggested that as the oceans and atmosphere warm, stronger storms, although not necessarily more storms overall, may be one result. As discussed in chapter 8, storm surges will reach higher on the shore as they are lifted by the higher sea level. As coastal managers develop plans to protect the shore from storms and storm surge, they will have to include the future elevation of sea level in their thinking.

Storms are the agents of much coastal change. Storm waves erode the shore, shape the beach, and move large quantities of sediment. The effectiveness of storms is determined by their intensities, paths, durations, and frequency. The elevated water level of a storm surge can cause significant damage, since it both allows waves to reach farther landward and can inundate a large area. Storm surge is responsible for most of the deaths during a hurricane. The combination of sea-level rise, climate change (which is likely to result in not only more, but more intense, hurricanes), and population growth along the coast poses substantial challenges for society.

10
EROSION AND SHORE PROTECTION

Most people's gut reaction to the erosion of beaches is that it must be stopped. But erosion is a natural phenomenon whose basic causes tend to be relatively simple: waves and currents, usually during sporadic storms. Ongoing sea-level rise also plays a role. As the water rises, the leading edge of the ocean moves landward, aggravating erosive processes. Storm waves chew the beach and eat the base of the foredune or backshore region. By excavating the area behind the beach, the storm waves make "new" sediment available to be carried by the longshore-drift system, which during a major storm operates at a very rapid rate. Every grain of sand on a beach was eroded from somewhere else and transported to the beach.

Although there can be no deposition without erosion, deposition is sometimes overlooked as a consequence of erosion. The material that is eroded from one place has to go somewhere else. Sand from a beach usually moves both alongshore and offshore. Fine-grained silts and clays can remain in suspension for a much longer time than sand and settle to the bottom in calmer areas often far away from the site of erosion. Deposition can cause its own set of problems. The sediments can fill navigation channels, necessitating dredging to maintain the channel or harbor. An influx of mud can smother shellfish beds and stifle marine vegetation.

Erosion becomes a "problem" only when it adversely affects human activities, so it is a social more than a geological concern. Any decision to combat erosion has at least two components: an understanding of the physical processes that cause the erosion in order to devise a strategy to fight it, and an assessment of the social and economic consequences of continued erosion in order to decide if the battle is worthwhile. If the cost of combating erosion exceeds the value, however determined, of whatever is saved, can the expense be justified?

An action taken within a given stretch of shoreline usually has consequences that extend beyond the immediate boundaries of the project. Erosion-control measures typically offer little or no protection from a major storm surge and virtually no defense from floods coming from upland waterways. The effects of the ongoing rise in sea level must be factored into the decision-making process as well.

Ownership of the shore may be a confusing factor. Property ownership in the coastal zone is not as simple as it might seem, and the regulations for land use and construction can be complex. One of the factors in the ownership of shorefront property is the definition of the seaward boundary, which varies by state (table 10.1). In some states, private ownership extends only as far as the high-tide line. In others, low tide is the limit. The circumstance that a tidal datum determines the seaward limit of private ownership further complicates the situation. As the shoreline retreats, the boundary moves along with it, and the area of the property decreases. The high- and low-tide lines move as sea level changes.

Like jetties at tidal inlets, any type of beach construction usually produces down-drift consequences. Thus all the owners of an affected stretch of shore should work together with a coherent plan, uniform methods and materials, and, perhaps, some type of cost sharing.

Often there are public benefits to combating shoreline erosion: the protection of infrastructure like roads, a decrease in siltation in navigation channels, and a limitation of silt that degrades water clarity and quality. The big question to address is whether to spend public funds to protect private property. In some situations, any effective plan for shore protection would be prohibitively expensive for individual landowners to undertake, but would be possible with public funding. Sometimes, government can justify projects on the ground of maintaining the tax

TABLE 10.1

Seaward Limit of Private Ownership

Mean high higher water	Mean high water	Mean low water
Texas	Alabama	Delaware
	Alaska	Massachusetts
	California	Maine
	Connecticut	New Hampshire
	Florida	Virginia
	Georgia	
	Louisiana	
	Maryland	
	Mississippi	
	New Jersey	
	New York	
	North Carolina	
	Oregon	
	Rhode Island	
	South Carolina	
	Washington	

Source: Data from National Research Council, Committee on National Needs for Coastal Mapping and Charting, Ocean Studies Board, Mapping Science Committee, Division of Earth and Life Sciences, *A Geospatial Framework for the Coastal Zone* (Washington, D.C.: National Academies Press, 2004), fig. 2.1.

base; if structures or land are lost to erosion, they cannot be taxed. But public expenditures should come with strings attached. The most common is the requirement that the use of public dollars for coastal development open the beach to meaningful public access. In addition to walkways leading to the beach, "meaningful access" can include reasonable parking areas.

When preserving or enhancing public beaches, questions can be complex and answers seldom are easy. What criteria must be met to justify the very real expenses? Every location has a unique mix of circumstances, and rarely does any one solution satisfy all parties.

* * *

In general, shoreline-protection measures are characterized as hard or soft, although a few methods might be somewhere in between. Hard methods are those in which a substantial structure is built to halt erosion: a seawall, riprap, groin, or breakwater. Soft methods, such as beach nourishment, tend more toward working around the problem, deflecting it rather than confronting it. The decision about what type of erosion protection, if any, is appropriate is, in part, a balance of the value of what is being protected and the cost of the protections (figure 10.1). There are four general strategies for dealing with shoreline erosion or the loss of shore as a result of sea-level rise:

- Spend what is necessary to protect the shore
- Accommodate the erosion, or "live with it"

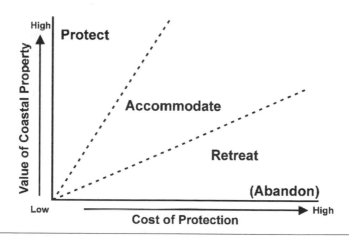

Figure 10.1 Management strategies for the consequences of shoreline erosion and sea-level rise. The nonmonetary aspects of the value of the property and the cost of protection, such as the historic or environmental significance of the property and the environmental change resulting from the protection, should be included in the decision-making process.

- Retreat from the shore by moving infrastructure and activities inland
- Abandon what is on the shore

A seawall is the epitome of hard shoreline-protection structures. A properly designed and constructed seawall stabilizes the shoreline, but it may not serve any other purpose. If it is built tall enough so that storm surge and waves will not overtop it (now or in the future, with higher sea level), is footed deep enough that scour will not undercut it, is constructed of materials that will not break under the onslaught of pounding waves, and is sufficiently tied into the land so that it will not topple, a seawall becomes, in effect, an artificial cliff that will stay fixed in space. Houses, other buildings, and roads behind such a seawall will be safe from destruction by waves and loss to erosion, but they will not be protected from flooding from behind. The presence of the structure will change or block the view of the beach. The cost of such a seawall is very high—hundreds, sometimes thousands, of dollars per foot—and there is no guarantee that a beach will remain in front of it. Indeed, in many instances, the beach disappears. Beach loss in front of a seawall generally is due to two factors. The seawall blocks the natural addition of new sediment from the backshore, and thus beach sand removed by longshore drift may not be easily or naturally replaced. In addition, the seawall reflects waves that hit it, and the interactions between the reflected and incident waves cause substantial turbulence and erosion. And with a poorly designed or constructed seawall, the loss of the beach can lead to the failure of the seawall itself. Figure 10.2 shows a seawall that collapsed at Sandbridge Beach on the Atlantic coast of southeastern Virginia.

Because the potential problems of seawalls are so well known, coastal engineers have developed methods that try to counteract them. A seawall can be built sufficiently far back from the shoreline so that storm waves are unlikely to strike it. This is problematic for at least two reasons. First, there may not be enough distance between the shoreline and whatever is to be protected; second, if you bet that storm waves will not reach the wall, and lose that bet, it is likely that you will lose the beach as well. Given that sea level is rising along most of the developed shore of the United States, it is probable that some time soon waves will reach the seawall.

Figure 10.2 The failure of the seawall at Sandbridge Beach, Virginia, in 1994. (Photograph used with permission of the Shoreline Studies Program, Virginia Institute of Marine Science, College of William & Mary)

Riprap, a sloping buffer of large rocks, properly designed and built in front of a seawall, can alleviate some of the pressure on the wall. By presenting a slanted, somewhat porous, many-faced body to the oncoming waves, riprap can dissipate the wave energy across a much greater surface area than that of the wall, by reflecting the waves in many directions. Some seashore plants may take root in and animals may dwell in the many nooks and crannies between the individual rocks of the riprap.

To facilitate the construction of a riprap revetment, relatively small rocks may be used as a core or base for the structure. The very large rocks needed to resist major waves are only an outer armor. In areas where sufficient quantities of big rocks are not readily available, engineers use concrete structures. There are a few proprietary shapes, often akin to gigantic toy "jacks." This general shape allows structural interlocking among adjacent modules, and the individual units can be sized for the local wave environment.

Another shore-protection option is the installation of riprap without the seawall behind it. Again, construction practices and design criteria are very important. As with vertical seawalls, riprap revetments block people's easy access to the water, and they must be designed for the anticipated rise in sea level.

In some locations, the shoreline has been hardened by the construction of a seawall or riprap in an attempt to stem the retreat of a bluff. Storm waves remove beach sediment and undercut the bluff, causing it to fail. When the cliff face collapses, sediment often falls in a slab-like mass, rebuilding the beach but moving back the crest of the bluff. The working hypothesis is that by building a revetment or seawall to prevent waves from attacking the base of the cliff, it will stop eroding. The hypothesis is only partially true. If properly designed and constructed, the structure that hardens the shoreline will keep storm waves from attacking and undercutting the bluff. But because the constant erosion has left the face of the bluff overly steep, the top will continue to erode and retreat until the slope is shallow enough to be stable. So cliff-top houses must be built well back from the edge. The natural transition of an unstable 70-degree slope to a more stable 30 degrees can take several decades. Unless some of the slump material can get to the water's edge, the beach will erode, too. As the waves and longshore currents continue to remove the beach, perhaps at a rate accelerated by wave reflection off the seawall, the beach will disappear as it loses the source of new sediment.

A major problem with seawalls and revetments is their "end effects." The lateral ends of these structures are points of great weakness. Because the shore adjacent to a seawall is not protected, it continues to erode and retreat. Thus the protected area sticks out into the water. If the ends of the structure do not turn and run back into the land, waves and longshore currents will attack the soil behind the seawall, weakening it and hastening its collapse. Because of the severity of end effects, it seldom makes sense to protect one house lot with a seawall while leaving adjacent properties unprotected. As noted earlier, protecting a long reach of shore with a single structure built to a uniform design almost always is more effective than a piecemeal approach. However, this strategy requires that all the landowners act together.

Historically, a groin has been a very common, hard shoreline-protection structure. The concept of a groin is very simple: construct a wall perpendicular to the coast, across the beach and well out into the surf zone to block the longshore transport of sediment. The groin disrupts the longshore current, so sediment accumulates on the up-drift side of the groin, causing the beach to build outward. The (unnaturally) widened beach is a protective buffer for the back-beach area. Unfortunately, it is not that simple.

When you look at a beach that has a groin, you will see one of the major problems of this tool: there seldom is just one groin. Because the waves that cause the longshore current are not disrupted, the current resumes just a short distance down-drift of the groin. And as no sediment moves across the groin, because it was trapped on the up-drift side, the longshore current now erodes sand from the shallow nearshore area down-drift of the structure. Thus it is necessary to construct another groin a little down-drift of the first to hold sand and limit the down-drift, erosional shadow of the first. So one groin usually is just the first in a series of them.

Engineers have developed formulas for the spacing between groins. Some have devised ways to reduce the trapping effect of groins by lowering their height so that some sand can pass over them, or they build them so that some sediment can pass through them. Other engineers have worked with the shape of the groins, making them L- or T- shaped, in order to influence wave transformation and minimize the down-drift shadow.

Groins are built of almost any imaginable material. Timber, steel sheet pile, and rock are the most common materials. Groins made from old refrigerators or automobile carcasses, segments of well casing or culvert suffer both aesthetically and functionally. Using cheaply available materials seldom works well. If you decide to use groins, the design must be robust enough for the local conditions, or the structure may fail and leave rubble to spoil the remaining beach.

An offshore breakwater is another type of structure that can alleviate erosion, although it is more often built to provide a protected area for an anchorage or a harbor. In effect, a breakwater is an artificial island that protects the shore under its lee. Classically, breakwaters are massive and are built an appreciable distance offshore.

The traditional breakwater is very costly, often prohibitively so. The offshore location requires that it be built from barges, and the water depth consumes an immense volume of stone. In navigable water, the finished structure has to be marked and must appear on nautical charts. Even if the breakwater works, there still can be problems. A breakwater

Figure 10.3 A gapped, nearshore breakwater system at Yorktown, Virginia, on the York River, a subestuary of Chesapeake Bay. The angled shape of the breakwater is designed to face into the two major directions from which waves approach this section of the coast. (Photograph by the author)

can make the wave conditions worse. Wave diffraction and refraction around an offshore breakwater can bring opposing waves together within the protected area, increasing rather than reducing their energy. A massive breakwater diminishes the quality of the ocean view. Many swimmers enjoy playing in the light and moderate waves that might otherwise reach the beach on a pleasant summer day without the breakwater.

A nearshore, or even shore-attached, breakwater in association with some beach fill is another means of coastal protection. Smaller breakwaters built close to shore are used to limit erosion on both ocean and estuarine coasts (figure 10.3). Depending on the local situation, relatively small, nearshore breakwaters can be designed as an individual unit protecting a short stretch of shore or as a series of units protecting a longer reach. By using data on the area's winds, coastal engineers can calculate the directions from which larger waves approach the shore. They feed data on storm winds and waves, storm surges, and different configurations for a breakwater into a wave-refraction model. Variables in breakwater design include distance offshore, length of individual elements, spacing between adjacent breakwaters, angle of the structure relative to the shoreline, and elevation of the top of the breakwater. The anticipated future sea level should be a design factor, too. Combining all this information, engineers can estimate the shape of the shoreline that would develop with each of the breakwater's configurations and, then, select the best design. Frequently, after building a nearshore breakwater, in order to shorten the time required for a shoreline to stabilize, sand is brought in and placed in a shape that approaches the contour that the shore eventually will assume. In some circumstances, sloping or grading the bluff behind the project not only provides the extra sand, but also stabilizes the bluff. The resulting shoreline behind the nearshore breakwater is curvy, with the beach extending outward behind the individual breakwaters and retreating into small embayments associated with the gaps. Small-scale wave diffraction and dispersion effectively limit the distance that the embayments can retreat into the mainland.

Nearshore breakwater systems function as artificial headlands, so there essentially is no erosion behind them. While waves push through the openings between individual breakwaters, they diffract and disperse in the protected water. The sand immediately behind the breakwaters becomes a *tombolo*, connecting the breakwater to the shore. The flanks of the tombolos increase the length of the shoreline across which the wave energy is spent, lessening the concentration of the energy. A system of gapped, nearshore breakwaters can protect a stretch of coast without the need to place riprap along its total length and, thus, is less costly. Also, because nearshore breakwaters are designed with the assumption that boats will not operate in their lee and are constructed in shallow water, they are not nearly as costly as the deeper-water breakwaters designed to protect harbors.

In recent decades, beach nourishment has become a common, soft method of protecting shorelines from erosion. About 900 individual projects have been undertaken along the Atlantic and Gulf coasts of the United States. As with many issues concerning beaches, the concept is fairly simple, but the application often is more complex. If you assume that erosion will remove a given quantity of sand from a beach each year, just put back the same quantity of sand that would be lost over the design life of the project. Beach nourishment actually is the building of a sacrificial beach. So embarking on a replenishment project embodies the understanding that it is not a one-time, permanent solution but an ongoing program with ongoing costs. And, as with all erosion-control projects, economics are very important.

Do the benefits of beach nourishment justify the expenses? Some of the costs are easy to compute: What is the contractor going to charge to obtain the sand and put it on the beach? But basic cost-benefit analysis often ignores the hidden, indirect costs: If trucks haul the sand, what is the wear on the highway system? What is the value of disrupting traffic? If each truck can carry 20 cubic yards (15 m³) of sand, and the project calls for 100,000 cubic yards (76,455 m³), that adds up to 5,000 truck trips. And 100,000 cubic yards is a small project. What are the environmental consequences of mining 100,000 cubic yards of sand? Dug to a depth of 10 feet (3 m), that would yield an opening about 520 feet (160 m) on a side, with an area a little greater than 6 acres (2.5 ha).

Large nourishment projects frequently rely on sand dredged from offshore and barged or pumped to a beach (figure 10.4). Projects using 1 million cubic yards (765,000 m³) of sand are common. A million cubic yards might seem like a frighteningly large volume, but it must be considered in perspective. To obtain that much sand, a square 1,000 yards (915 m) on a side would have to be mined 3 feet (1 m) deep. That is an area of about 200 acres (80 ha)—again, a large number. But the standard nautical chart of nearshore regions has a scale of 1:80,000; thus the 200-acre dredging site would appear as a square with sides just under 0.5 inch (1 cm). On the more detailed, 1:40,000 scale charts, that site would be less than 1 inch square (2.3 cm²). In that perspective, the mined area does not seem large, especially when seen on charts that commonly measure 30 inches (75 cm) or more on a side.

Figure 10.4 Sand being pumped onto the beach at Garden City, South Carolina, as part of a nourishment project. (Photograph by the author)

There is a very rough rule of thumb for beach nourishment projects that 1 cubic yard (0.8 m³) of beach fill will widen 1 foot (30 cm) of beach by 1 foot. Stated another way, 9 cubic yards (7 m³) of fill will widen 1 yard (1 m) of beach by 1 yard. So increasing the width of 1,000 feet (300 m) of shorefront by 60 feet (20 m) would call for 60,000 cubic yards (46,000 m³) of material. This probably is a low estimate, as the sand redistributes relatively quickly to a more stable profile.

Mining sand offshore has real environmental consequences. We must assume that every worm and small, shelled animal living in the sand that is dredged will die. Fortunately, the clean, coarse sands preferred for beach nourishment tend to have a low density of in-dwelling animals, and those species tend to recolonize barren areas relatively quickly. But there are other consequences to consider—for example, the impact of the noise of the underwater dredging on marine mammals like whales and dolphins, the effect of the short-term loss of potential habitat on fish that feed on the worms and may make only seasonal use of the area, and the possibility of catching endangered sea turtles in the dredge. Many of these potential hazards can be managed, but not eliminated, by limiting the time of year when the mining takes place and suspending operations when full-time spotters sight an endangered species.

Again, the question to address for each beach-nourishment project is: Do the benefits offset the costs? But what are the benefits? Each potential project must be analyzed on its own terms. A successful program will protect infrastructure from the effects of beach erosion; thus one economic benefit is the value of the roads and buildings immediately behind the beach that may be lost or damaged in a storm that has a reasonable likelihood of occurring during the project's anticipated life. A wide, properly nourished beach is more aesthetically appealing to many people than a narrow, sand-starved beach and can be the backbone of a substantial tourist economy. The hotels, restaurants, and T-shirt shops that draw beach-loving tourists all benefit from the beach-protection project.

It is difficult to estimate when a nourishment project will need supplemental work. If an unusually severe hurricane or nor'easter should strike early in the project's planned life, critics, understandably, may complain about the design and the cost of the unanticipated renourishment. But if the planned life had been 10 years and the storm was a "50-year" storm, was the planning in error? Conversely, supporters may claim an unexpectedly rapid and high "return on investment," asserting that the nourished beach saved the value of infrastructure from the storm. Project life may be as much a sociopolitical as a physical problem. Engineers and scientists sometimes do not present information so that the general public can find it or understand it. But working with nature requires making assumptions that include various levels of confidence and ranges of error. Proponents of a specific nourishment program should not oversell an optimistic project life.

Sea-level rise must be part of the planning for a beach-nourishment project. Rising sea level will affect not just the immediate area of the replenishment but the entire region. There may be no point in nourishing a barrier-island beach if the island eventually will be inundated from behind by the rising water in the back-barrier lagoon. If sea-level rise will render an area less habitable or completely uninhabitable, there may be no need to restore a beach.

The sand used in a nourishment project may be the single most important design element. Although it should be obvious, it often is necessary to state that the sand must not be polluted. Even if it could be demonstrated that the sand would wash clean of contaminants almost instantly, it is unlikely that the public would "trust" the new beach. Fortunately, sand usually is clean because it is more porous and permeable than silts and clays, which tend to hold contaminants or pollutants more tenaciously than sand.

The sand has to be reasonably available. No matter if engineers determine the most compatible grain-size characteristics for a nourishment project, if sand that meets those specifications is not accessible in a sufficient quantity, they will have to reevaluate the project and adjust the design for a less-than-optimal sediment.

The general rule is that the sand placed on the beach should have grain-size characteristics as close as possible to those of the natural beach, with a slight preference for the new material being a little coarser. Color is important, too, with a preference again toward matching the existing sediment or using white, light tan, or even light pink sand. However, color is difficult to judge. Frequently, as a nourishment project is in progress, people are dissatisfied with the color of the new beach, but the disappointing dark gray sand of some recently nourished beaches often bleaches to a much more acceptable tan. The color change, which can take a year or more, results from exposure to the sun and oxidation.

With beach-nourishment projects, bigger tends to be better. Inevitably, a newly replenished beach loses sand, both along the front of work and at the edges. The farther apart the edges—that is, the greater the alongshore length of the nourished area—the less significant the end losses when compared with the entire sand volume of the project. Sand redistributed away from the nourished beach can widen and enhance the adjacent beaches, spreading the benefit of the project beyond the explic-

Figure 10.5 The northern end of Assateague Island National Seashore, November 1999. The jetties at the inlet at Ocean City, Maryland (at the top of the photograph), have disrupted the longshore transport system, so the up-drift portion of Assateague Island is sand starved and is eroding very rapidly. The National Park Service has since nourished part of the retreating area. (From U.S. Army Corps of Engineers, U.S. Army Engineer Research and Development Center, Coastal Inlets Research Program, Inlets Online, http://www.ocean science.net/inletsonline)

itly replenished area. However, the new sand also can wind up in places, such as dredged channels, where it may not be desirable.

Beach-nourishment projects can abate problems that rise from other actions. A recent example is the restoration undertaken on Assateague Island, immediately down-drift from the jettied inlet at Ocean City, Maryland (figure 10.5). Although most of Assateague is a national seashore and the National Park Service traditionally avoids interfering with natural processes, the Park Service made the decision to nourish the island. The long jetties at Ocean City Inlet have trapped sand that would have moved alongshore to Assateague Island. As a result, the island's northern end is losing critical wildlife habitat, and the mainland risks direct attack by ocean waves. Thus the Park Service made the judgment that beach nourishment compensates for the disruption to the natural sand supply that the longshore current should provide. Some beach-nourishment projects in Florida also make up for problems caused by "controlling" inlets. As such, nourishment is a benign method of shore protection, since it mimics the natural processes that existed before the construction of jetties.

* * *

Accommodating, or learning to live with, erosion and sea-level rise along the beach presents another set of issues. Structures can be built on pilings, allowing storm waters to flow beneath them. Local or state regulations can mandate that new construction be set back some distance from the waterline. This is a form of retreating from the eroding shore. Depending on the state and locality, the distance of the required setback may be a specific distance—for example, 100 feet (30 m)—or may be based on the historical rate of erosion or another factor, such as the type of land behind the beach. If the setback would put the building on a coastal dune or in another critical habitat, it most likely will not be possible to build. With or without a setback, as the shoreline moves toward the landward boundary of the property, the remaining building lot eventually will become too small for construction. The landowner will have the sorrow of losing an opportunity to live by the shore, and the locality will lose the tax revenue of a valuable, coastal property.

Most people who want to build along the shore or buy a beach house have to borrow money to help pay for it. Lending agencies require that the applicants buy flood insurance to cover potential losses. Several insurance companies have stopped writing policies for many coastal areas because the probability of loss is so great. The National Flood Insurance Program, administered by the Federal Emergency Management Agency (FEMA), enables property owners to purchase insurance to protect against losses from flooding, including coastal flooding, when it is not available from commercial insurers. The program is controversial because many people believe that it facilitates construction in areas where flooding and major damage are very likely; without the subsidized flood insurance, there would be no loan to build or buy the beach house.

The insurance companies and lenders, in order to minimize their exposure to potential losses, may require specific construction practices. Even then, the premiums may be very high. If an insurer believes that an area is likely to be hit by a major hurricane within a decade, the annual premium for a coastal property would be for at least one-tenth of its value. If the history of losses in a coastal area is great, the minimum "deductible" expense may be high.

* * *

Shoreline erosion is a natural phenomenon that itself is not a problem. It becomes so when it has negative impacts on human activities. Shoreline managers should attempt to stem erosion only after considering many factors: What will be protected? What is the "value" of the protected natural or constructed features? What are the potential detrimental consequences of the action? What is the cost of the contemplated action?

Seawalls and revetments, if properly designed and constructed, usually can stop shoreline retreat but frequently at the loss of the beach. Groins can successfully trap sand, but they simultaneously starve adjacent areas. Beach nourishment can protect the beach but require periodic replenishment.

Protecting the shoreline is not always the best response to beach erosion. Sometimes forgoing expensive development is the wisest option. Providing basic access to a beach with a minimum amount of infrastructure (dressing and sanitary facilities) can be the best course of action. Every beach has its own physical character, and the solutions are complex.

APPENDIX 1: UNITS OF SPEED
Approximate Conversion Factors

1 mile per hour (common, "statute" mile)
- 88 feet per minute
- 1.6 kilometers per hour
- 1.5 feet per second
- 0.87 knot
- 0.45 meter per second

1 knot
- 1 nautical mile per hour
- 101 feet per minute
- 1.85 kilometers per hour
- 1.7 feet per second
- 1.15 miles per hour
- 0.5 meter per second

1 meter per second
- 197 feet per minute
- 3.6 kilometers per hour
- 3.3 feet per second
- 2.2 miles per hour
- 1.9 knots

1 kilometer per hour

 55 feet per minute

 0.9 feet (11 inches) per second

 0.62 mile per hour

 0.54 knot

 28 centimeters per second

APPENDIX 2: BEAUFORT SCALE

Beaufort number	Knots	Seaman's terms	World Meteorological Organization (1964)	Observations at sea	Observations on land
0	<1	Calm	Calm	Sea like mirror	Calm; smoke rises vertically
1	1–3	Light air	Light air	Ripples with appearance of scales; no foam crests	Smoke drift indicates wind direction; vanes do not move
2	4–6	Light breeze	Light breeze	Small wavelets; crests of glassy appearance; not breaking	Wind felt on face; leaves rustle; vanes begin to move
3	7–10	Gentle breeze	Gentle breeze	Large wavelets; crests beginning to break; scattered whitecaps	Leaves and small twigs in constant motion; light flags extended
4	11–16	Moderate breeze	Moderate breeze	Small waves, becoming longer; numerous whitecaps	Dust, leaves, and loose paper raised up; small branches move
5	17–21	Fresh breeze	Fresh breeze	Moderate waves, taking longer form; many whitecaps; some spray	Small trees in leaf begin to sway
6	22–27	Strong breeze	Strong breeze	Large waves forming; whitecaps everywhere; more spray	Larger branches of trees in motion; whistling heard in wires
7	28–33	Moderate gale	Near gale	Sea heaps up; white foam from breaking waves begins to be blown in streaks	Whole trees in motion; resistance felt in walking against wind

Beaufort number	Knots	Seaman's terms	World Meteorological Organization (1964)	Observations at sea	Observations on land
8	34–40	Fresh gale	Gale	Moderately high waves of greater length; edges of crests begin to break into spindrift; foam is blown in well-marked streaks	Twigs and small branches broken off trees; progress generally impeded
9	41–47	Strong gale	Strong gale	High waves; sea begins to roll; dense streaks of foam; spray may reduce visibility	Slight structural damage occurs; slate blown from roofs
10	48–55	Whole gale	Storm	Very high waves with overhanging crests; sea takes white appearance as foam is blown in very dense streaks; rolling is heavy and visibility reduced	Seldom experienced on land; trees broken or uprooted; considerable structural damage
11	56–63	Storm	Violent storm	Exceptionally high waves; sea covered with white foam patches; visibility still more reduced	Very rarely experienced on land; usually accompanied by widespread damage
12	64–71	Hurricane	Hurricane	Air filled with foam; sea completely white with driving spray; visibility greatly reduced	
13	72–80				
14	81–89				
15	90–99				
16	100–108				
17	109–118				

Source: N. Bowditch, *The American Practical Navigator* [*Bowditch*]. H. O. Publication, no. 9 (Washington, D.C.: Government Printing Office, 1966).

APPENDIX 3: SAFFIR-SIMPSON HURRICANE SCALE

Category	Definition			Storm surge		Effects
	mph	knots	km/hr	feet	meters	
1	74–95	64–82	119–153	4–5	1.2–1.5	No real damage to buildings and structures. Damage primarily to unanchored mobile homes, shrubbery, and trees. Some damage to poorly constructed signs. Also, some flooding of coastal roads and minor damage to piers.
2	96–110	83–95	154–177	6–8	1.8–2.4	Some roofing-material, door, and window damage to buildings. Considerable damage to mobile homes, poorly constructed signs, and piers. Coastal and low-lying escape routes flood 2–4 hours before arrival of the center of the hurricane. Small craft in unprotected anchorages break moorings.
3	111–130	96–113	178–209	9–12	2.7–3.7	Some structural damage to small residences and utility buildings, with a minor amount of curtain-wall failure. Damage to shrubbery and trees: with foliage blown off trees and large trees blown down. Mobile homes and poorly constructed signs are destroyed. Low-lying escape routes flooding near the coast 3–5 hours before arrival of the center of the hurricane. Flood destroys smaller structures, with large structures damaged by battering from floating debris. Terrain continuously lower than 5 feet (1.5 m) above mean sea level may be flooded inland 8 miles (13 km) or more. Evacuation of low-lying residences within several blocks of the shoreline may be required.

Category	Definition			Storm surge		Effects
	mph	knots	km/hr	feet	meters	
4	131–155	114–135	210–249	13–18	3.9–5.5	More extensive curtain-wall failures, with some complete roof structure failures on small residences. Shrubs, trees, and all signs are blown down. Complete destruction of mobile homes. Extensive damage to doors and windows. Low-lying escape routes may flood 3–5 hours before the arrival of the center of the hurricane. Major damage to lower floors of structures near the shore. Terrain lower than 10 feet (3 m) above sea level may be flooded, requiring massive evacuation of residential areas as far inland as 6 miles (10 km).
5	>155	>135	>249	>18	>5.5	Complete roof failure on many residences and industrial buildings. Some complete building failures, with small utility buildings blown over or away. All shrubs, trees, and signs blown down. Complete destruction of mobile homes. Severe and extensive window damage. Low-lying escape routes flooded 3–5 hours before arrival of the center of the hurricane. Major damage to lower floors of all structures located less than 15 feet (4.5 m) above sea level within 500 yards (460 m) of the shoreline. Massive evacuation of residential areas on low ground within 5–10 miles (8–16 km) of the shoreline may be required.

Source: After National Weather Service, National Hurricane Center, The Saffir-Simpson Hurricane Wind Scale Summary Table, http://www.nhc.noaa.gov/sshws_table .shtml?large.

GLOSSARY

ACCRETION The accumulation of sediment; the opposite of erosion.

AEOLIAN Wind blown.

ALBEDO The ratio of incident light to reflected light.

ALLOCHEMICAL Chemically formed sedimentary rocks that show evidence of transportation in the basin of their deposition—for example, shell hash.

AMPHIDROMIC POINT A location on a tide chart where the tidal range approaches zero and from which co-tidal lines radiate. The tidal cycle rotates around an amphidromic point.

AMPHIDROMIC SYSTEM The common tidal system in which the tidal cycle progresses in an orderly manner around a central point.

ANGLE OF REPOSE The steepest angle, measured from horizontal, at which loose, noncohesive sediments will come to rest when piled with similar material.

APHELION The point in its orbit around the sun where a celestial body is farthest from the sun.

APOGEE The point in its orbit where the moon is farthest from Earth.

AUTOCOMPACTION The compression of marsh peat under its own weight.

BARCHAN A crescent-shaped sand dune whose outer edges, or "horns," point down-wind.

BEACH CREST The break in slope that separates the beach face from the berm.

BEACH CUSP One of a series of regularly spaced, concave arcs on the beach face.

BEACH FACE The surface of a beach that slopes down from the berm to the water.

BEACH STEP See STEP.

BED LOAD The portion of moving sediment that is very near the bottom of a current, of either air or water, and that is not suspended in the flow.

BERM The relatively flat, terrace-like upper surface of the beach; it is the most hospitable part of the beach for sunbathing and playing.

BERM CREST *See* BEACH CREST.

BLOWOUT A trough-shaped depression formed by wind in a dune or another area of loose sand.

BOUNDARY LAYER The thin region close to a solid in a fluid where there is a rapid increase in velocity from near zero where the fluid is in contact with the solid to velocity of the unhindered flow.

CLAPOTIS A standing wave that results from the interaction of an incoming wave and a wave reflected off a vertical surface, such as a seawall.

CLOSURE DEPTH The depth below which there is essentially no change in the beach profile over time and no movement of sediment.

COHESIVE SEDIMENT Clay-bearing sediments in which the grains stick together when wet.

COQUINA A sedimentary rock that consists of fragments of shells.

CO-RANGE LINE The lines of equal tidal range on a tide chart.

CO-TIDAL LINE The lines on a tide chart that depict the places with the same tidal stage, such as high tide, occur at the same time.

CURRENT RIPPLES A repetitive pattern of ridges and troughs in sediment caused by a current; they are asymmetrical with the step side on the down-current side of the ridge.

CUSP *See* BEACH CUSP.

DEEP-WATER WAVE A wave in water deeper than one-half of its length.

DIFFRACTION The apparent bending of a wave as it passes an obstacle; the spreading of a wave as it passes through an inlet.

DISSIPATIVE BEACH A beach usually with a narrow berm and a wide, very shallowly sloping beach face and few or no offshore bars. Frequently, the shape of a beach immediately after a storm.

DIURNAL INEQUALITY The difference in elevation between the two successive high (or low) tides in a semi-diurnal tidal system.

DIURNAL TIDE A tidal system with one high and one low tide a day.

EDGE WAVE A low-frequency (infra-gravity) wave that runs parallel to the beach and decreases in height with distance offshore; it forms as a consequence of (wind) waves breaking.

EKMAN SPIRAL The rotation of a wind-generated current with increasing water depth.

EKMAN TRANSPORT The movement of water to the right (in the Northern Hemisphere) of the direction of the wind blowing across it. The surface water moves about 45 degrees to wind, but the angle increases with water depth, so the overall net transport of water is at a right angle to the wind. *See also* EKMAN SPIRAL.

EQUILIBRIUM BEACH A theoretical beach profile that is steeper near the shore and closely approaches horizontal farther offshore, with a gently concave shape that fits a specific mathematical formula. The equilibrium profile supposedly balances the sediment deposited and the sediment removed by waves and currents.

ESCARPMENT *See* SCARP.

EUSTATIC SEA LEVEL (EUSTACY) The global sea level, which varies with changes in the volume of water in the oceans and in the size and shape of the ocean basins without regard to local uplift or subsidence of land.

EXTRATROPICAL STORM A cyclone that draws most of its energy from the temperature contrast between warm and cold air masses. *See also* TROPICAL STORM.

FELDSPAR Any of a common group of aluminum-silicate minerals whose grains frequently constitute much of the sediment in beach sand.

FETCH The over-water length across which wind blows.

FLOTSAM Refuse or trash that is floating on the water. *See also* JETSAM.

FOREDUNE The line of sand dunes closest to the beach.

FORESHORE *See* BEACH FACE.

FREQUENCY The number of waves that pass a point in a fixed interval of time; the inverse of period.

GEOID The shape of Earth if the surface of the sea were continuous through the continents. A theoretical surface that is everywhere to the direction of gravity, it is the reference for several types of surveying.

GRAVITY WAVE A wave within which the primary force restoring the water to a flat surface is gravity. Most waves on the surface of the ocean are gravity waves. *See also* WIND WAVE.

GROIN (GROYNE) A shore-protection structure built perpendicular to the shoreline to trap sediment moving in the longshore transport system.

GROUNDWATER Water that is in the ground, as opposed to the surface water of rivers, lakes, and the ocean.

GROUP VELOCITY The speed at which a group of waves, a wave train, moves; one-half the speed of a solitary wave.

HABOOB An intense dust storm that occurs in arid regions.

HEAVY MINERAL A mineral with a density, or specific gravity, greater than 2.85.

INFRAGRAVITY WAVE A wave—such as a tide, a tsunami, and an edge wave—with a period longer than 20 seconds.

INVERTED BAROMETER EFFECT The elevation of the sea surface beneath an area of low atmospheric pressure.

ISOBATH A contour line that connects points of equal water depth, as often depicted on a nautical chart.

ISOSTASY Adjustment of the crust of Earth in response to changes in the load it places on the mantle. If the load increases—for example, by the formation of an ice cap—the crust would be pushed down into the mantle.

JETSAM Materials that are deliberately thrown (jettisoned) from a vessel, usually to lighten it during an emergency. *See also* FLOTSAM.

JETTY A structure similar to a groin built at an inlet to stabilize the location of the inlet.

LADDERBACK RIPPLES A hierarchal system of ripples in sediment, commonly occurring in the trough (runnel) between a nearshore ridge and the beach face where a set of small current ripples runs along the trough of a larger set of ripples.

LOESS A large, blanket-like deposit of fine-grained sediment, usually silt, that is generally believed to be an aeolian deposit.

LONGSHORE CURRENT A flow parallel to the beach formed by waves striking the beach at an angle.

LOW-TIDE TERRACE The usually wide, flat area immediately seaward of the beach face and step that is exposed only during very low tides.

MANTLE The viscous or plastic-like layer in the interior of Earth that is beneath the crust and above the core. It is separated from the crust by the Mohorovičić discontinuity, or Moho.

MEAN HIGHER HIGH WATER (MHHW) The average height computed for a tidal epoch of the higher of the two daily high waters at a location with a semi-diurnal or mixed tide.

MEAN HIGH WATER (MHW) The average height of all high waters recorded during a tidal epoch.

MEAN HIGH WATER NEAPS (MHWN) The average height of all high waters occurring during quadrature over a tidal epoch.

MEAN HIGH WATER SPRINGS (MHWS) The average height of all high waters occurring during syzygy over a tidal epoch.

MEAN LOWER LOW WATER (MLLW) The average height computed for a tidal epoch of the lower of the two daily low waters at a location with a semi-diurnal or mixed tide.

MEAN LOW WATER (MLW) The average height of all low waters recorded during a tidal epoch.

MEAN LOW WATER NEAPS (MLWN) The average height of all low waters occurring during quadrature over a tidal epoch.

MEAN LOW WATER SPRINGS (MLWS) The average height of all low waters occurring during syzygy over a tidal epoch.

MEAN SEA LEVEL (MSL) The average elevation of the water surface at all stages of the tide from hourly measurements over a tidal epoch.

MEAN TIDE LEVEL (MTL) The elevation of the water surface midway between mean high water and mean low water.

MEDANO A generally cone-shaped sand dune formed by winds blowing from several directions.

MIXED TIDE A tide similar to a semi-diurnal tide with two daily highs and lows, but with a great diurnal inequality.

NEAP TIDE The tide with a reduced range that occurs during quadrature.

NONCOHESIVE SEDIMENT Sediment, such as sand, in which the grains do not stick together.

NOR'EASTER (NORTHEASTER) A storm of the Middle Atlantic and New England coast that derives its name from strong northeast winds.

OOLITH A small, nearly spherical sedimentary particle usually formed by chemical precipitation in shallow water, most commonly composed of calcium carbonate.

ORTHOCHEMICAL Chemically precipitated sedimentary rocks that show little evidence of transportation.

OVERWASH Sand that was transported well inland from the beach by storm waves.

PARABOLIC DUNE A concave sand dune whose outer edges, or "horns," point up-wind.

PAVEMENT A surface of rocks or shell from which any loose sediment has been blown away. Along the coast, pavements often occur on the landward-most portion of a berm or in low areas between dunes.

PEAT An unconsolidated deposit of partially decomposed plant remains that forms in a wet environment such as a marsh; an early stage in the development of coal.

PERIGEE The point in its orbit where the moon is closest to Earth.

PERIHELION The point in its orbit around the sun where a celestial body is closest to the sun.

PERIOD The number of seconds it takes for one wave to pass a point; the inverse of frequency.

PERMEABILITY The ability of a porous rock to allow the transmission of fluid.

POROSITY The ratio of the empty, or void, space to the total volume of a rock or sedimentary deposit.

PRIMARY DUNE See FOREDUNE.

PROVENANCE The source area of sediments.

QUADRATURE The configuration in which the sun and the moon are at a right angle from each other, as seen from Earth, giving rise to neap tides. The moon is in its first or last quarter.

REFLECTIVE BEACH A beach with a relatively steep beach face and often a full berm, as would occur after an interval of sustained beach growth or accretion.

REFRACTION The apparent bending of waves as wave velocity decreases in shallowing water.

RELATIVE DENSITY *See* SPECIFIC GRAVITY.

RELATIVE SEA LEVEL The level of the sea at a location. Changes in relative sea level are the combination of worldwide, eustatic changes and local changes of whatever cause.

RIDGE-AND-RUNNEL SYSTEM A configuration of a shore-parallel bar, or ridge, and a trough, or runnel, between the ridge and the beach face, such that the ridge moves through the runnel and attaches, or welds, to the beach face, enlarging the beach.

RIP CURRENT A usually narrow flow of water perpendicular to the beach that passes through the surf zone.

RIPRAP A construction of rocks designed to protect the shore from the erosive action of waves; the rocks used in the construction.

SCARP A small cliff. In coastal geology, a small cliff in either the beach of face of the primary dune usually cut by wave erosion during a storm.

SEA The set of waves with differing wave lengths in the area of their generation.

SEAWALL A rigid, vertical face constructed parallel to the shoreline to resist erosion.

SEBKHA A salt flat; an area in a hot and arid setting that is occasionally flooded at very high tides, after which the water evaporates, leaving a salt crust.

SEICHE A wave that moves back and forth across a confined or partially confined body of water, sometimes described as the "bathtub effect."

SEMI-DIURNAL TIDE A tidal system with two high and two low tides a day.

SHALLOW-WATER WAVE A wave in water shallower than one-twentieth of its length.

SHORELINE The line where the beach and water surfaces meet.

SIDEREAL MONTH The time required for the moon to complete a full orbit with reference to the stars around Earth: approximately 27 days, 7 hours, and 43 minutes.

SLIPFACE The steep side of a current ripple or sand dune that advances down-current.

SPECIFIC GRAVITY The density (weight per unit volume) of a substance compared with the density of water.

SPRING TIDE The tide with an increased range that occurs during syzygy.

STEP The small notch at the bottom of the beach face that is commonly composed of coarser sediments than those that occur landward or seaward.

STORM BEACH *See* DISSIPATIVE BEACH.

STORM SURGE The difference between the astronomical (predicted) and the observed water levels, that difference resulting from the effects of weather.

STORM TIDE The observed water level during a storm.

STRATIFICATION The occurrence or formation of layers. In water, stratification usually results from the difference in density of two water masses; that difference can be a due differences in temperature or salinity.

SURF BEAT The low-frequency variations in nearshore water level caused by the interaction of sets of waves with different frequencies.

SWASH BAR A nearshore (sand)bar built up to water level by wave action; a ridge in a ridge-and-runnel system.

SWELL A series of regular, evenly spaced wind waves that have traveled from their area of generation.

SYNODIC MONTH The time required for the moon to complete a full orbit with reference to the sun around Earth: approximately 29 days, 12 hours, and 44 minutes.

SYZYGY The configuration in which the sun, moon, and Earth are aligned with one another, giving rise to spring tides. The moon is either new or full.

TERRIGENOUS Derived from Earth's surface.

THERMOCLINE The region of very rapid temperature change. In the ocean, it usually separates warmer surface waters that are mixed by wave action from cooler, deeper waters.

TIDAL EPOCH The 19-year period of observations used for the determination of tidal datums; the 18.61-year cycle for the repetition of tide-producing forces.

TIDAL PRISM In an estuary or a lagoon, the difference in water volume between high and low tide; the volume of water flowing through a tidal inlet during the change between high and low tides.

TIDAL RANGE The difference in elevation of the water surface at the times of consecutive high and low tides.

TOMBOLO A sand or gravel bar extending from the mainland toward an island.

TRANSGRESSION The landward movement of the shoreline during a period of rising sea level.

TRANSVERSE DUNE A sand dune elongated across the direction of the prevailing wind and generally formed in an area with little vegetation.

TROCHOID The curve traced by a point on a radius of a circle that rolls without slipping on a straight line; the shape of a wave.

TROPICAL STORM A cyclone, including hurricanes, that draws most of its energy from the latent heat of condensation (the energy released when water vapor changes to liquid). *See also* EXTRATROPICAL STORM.

TSUNAMI A long-period wave formed by a sudden disturbance to the seafloor, such as an earthquake.

UPWELLING The rising of cold water to the surface along a coast in response to the offshore flow of warm surface water due to Ekman transport.

VENTIFACT A rock or pebble abraded by wind-blown sand.

WAVE BASE The maximum depth at which waves can initiate movement of bottom sediment.

WAVE LENGTH The linear distance between corresponding parts of consecutive waves; the distance between successive wave crests.

WAVE RIPPLES A long crested, symmetrical ripple mark formed in surface sediments by the oscillatory water motion under waves.

WAVE TRAIN A group of waves with the same period or wavelength.

WIND WAVE A surface wave generated by wind. *See also* GRAVITY WAVE.

BIBLIOGRAPHY

1. BEACHES

Bentz, T. One small tern deserves another. September 1998. Smithsonian Institution, Migratory Bird Center. http://nationalzoo.si.edu/scbi/migratorybirds/featured _birds/default.cfm?bird=Least_Tern(accessed August 15, 2011).

Carson, R. *The Edge of the Sea*. Boston: Houghton Mifflin, 1955.

Davis, R. A., Jr., and D. M. FitzGerald. *Beaches and Coasts*. Malden, Mass.: Blackwell Science, 2004.

Dean, R.G.. *Beach Nourishment: Theory and Practice*. Advanced Series on Ocean Engineering, vol. 18. River Edge, N.J.: World Scientific, 2002.

Hobbs, C. H., III, C. B. Landry, and J. E. Perry III. Assessing anthropogenic and natural impacts on ghost crabs (*Ocypode quadrata*) at Cape Hatteras National Seashore, North Carolina. *Journal of Coastal Research* 2, no. 6 (2008): 150–158.

Morang, A., and L. E. Parson. Coastal terminology and geologic environments. In U.S. Army Corps of Engineers, *Coastal Engineering Manual*, EM 1110–2-1100, part 4, chap. 1. 2002. http://chl.erdc.usace.army.mil/CHL.aspx?p=s&a=ARTICLES;104 (accessed August 8, 2011).

National Oceanographic and Atmospheric Administration. NOAA aircraft takes dramatic photos of North Carolina coast after Hurricane Isabel unleashed her fury. September 23, 2003. NOAA News Online. http://www.noaanews.noaa.gov/stories/ s2091.htm; http://www.noaanews.noaa.gov/stories/images/hatteras-new-inlet-comparison.jpg (accessed August 8, 2011).

Neal, W. J., O. H. Pilkey, and J. T. Kelley. *Atlantic Coast Beaches: A Guide to Ripples, Dunes, and Other Natural Features of the Seashore*. Missoula, Mont.: Mountain Press, 2007.

North Carolina Coastal Federation. Cape Hatteras Coast Keeper issues: Hurricane Isabel damage report. http://www.nccoast.org/ch-issues.htm (accessed February 25, 2005).

Pilkey, O. H. *A Celebration of the World's Barrier Islands.* New York: Columbia University Press, 2003.

Shumway, S. W. *The Naturalist's Guide to the Atlantic Seashore: Beach Ecology from the Gulf of Maine to Cape Hatteras.* Guilford, Conn.: Globe Pequot Press / Falcon Guide, 2008.

Stewart, R. H. Coastal processes and tides. In *Introduction to Physical Oceanography.* 2005. Texas A&M University. http://oceanworld.tamu.edu/resources/ocng_textbook/chapter17/chapter17_01.htm (accessed August 8, 2011).

U.S. Army Corps of Engineers. Coastal morphodynamics. In *Engineering and Design: Coastal Geology.* Coastal Engineering Manual, EM 1110–2-1810, chap. 4. 1995. http://140.194.76.129/publications/eng-manuals/em1110-2-1810/c-4.pdf (accessed August 8, 2011).

——. Hurricane Isabel breaches Hatteras Island. Engineer update, 27(11). 2003. http://www.hq.usace.army.mil/cepa/pubs/nov03/story8.htm (accessed February 25, 2005).

——, U.S. Army Engineer Research and Development Center, Coastal Inlets Research Program. Inlets Online. http://www.oceanscience.net/inletsonline/ (accessed August 8, 2011).

U.S. Fish and Wildlife Service. The Atlantic Coast piping plover. 2007. http://www.fws.gov/northeast/pipingplover/pdf/plover.pdf (accessed August 15, 2011).

A Virginia Gazetteer. University of Virginia Library. http://fisher.lib.virginia.edu/collections/gis/vagaz/search_by_quad.php (accessed January 4, 2010).

Welland, M. *Sand: The Never-Ending Story.* Berkeley: University of California Press, 2009.

Wright, L. D. *Morphodynamics of Inner Continental Shelves.* Boca Raton, Fla.: CRC Press, 1995.

Wright, L. D., and A. D. Short. Morphodynamic variability of surf zones and beaches: A synthesis. *Marine Geology* 56 (1984): 93–118.

2. WIND

Baker, B. B., Jr., W. R. Deebel, and R. D. Geisenderfer, eds. *Glossary of Oceanographic Terms.* 2nd ed. U.S. Naval Oceanographic Office Special Publication 35. Washington, D.C.: Government Printing Office, 1966.

Bowditch, N. *The American Practical Navigator [Bowditch].* H. O. Publication, no. 9. Washington, D.C.: Government Printing Office, 1966.

Hurler, S. *Defining the Wind: The Beaufort Scale and How a Nineteenth-Century Admiral Turned Science into Poetry.* New York: Crown, 2004.

National Weather Service, National Hurricane Center. The Saphir-Simpson Hurricane Wind Scale. Last modified August 11, 2011. http://www.nhc.noaa.gov/aboutsshs.shtml (accessed August 15, 2011); http://www.nhc.noaa.gov/sshws_table.shtml?large (accessed September 14, 2011).

3. WAVES

Anthoni, J. F. Oceanography: Waves: Theory and principles of waves, how they work and what causes them. 2000. http://www.seafriends.org.nz/oceano/waves.htm (accessed January 7, 2010).

Bascom, W. *Waves and Beaches*. Garden City, N.Y.: Doubleday / Anchor Books, 1980.

Bowditch, N. *The American Practical Navigator* [*Bowditch*]. H. O. Publication, no. 9. Washington, D.C.: Government Printing Office, 1966.

Kearney, M. S. Geography 140, Coastal environments. 2004. University of Maryland. http://www.geog.umd.edu/homepage/courses/140/ppt/08%20Types%20of%20Waves.ppt (accessed June 3, 2005).

Newman, G., and W. J. Pierson Jr. *Principles of Physical Oceanography*. Englewood Cliffs, N.J.: Prentice-Hall, 1966.

Schlicke, T. Breaking waves and the dispersion of surface films. Ph.D. diss., University of Edinburgh, 2002. http://www.ph.ed.ac.uk/~ted/thesis/node10.html (accessed June 3, 2005).

Smith, N. P. *A Qualitative Discussion of Ocean Wind Waves*. Contribution, no. 7. Madison: Marine Studies Center, University of Wisconsin, 1971.

Texas Parks and Wildlife. Texas Gems. http://www.tpwd.state.tx.us/landwater/water/conservation/txgems/lagmadr/index.phtml (accessed September 14, 2011).

4. TIDES

Baker, B. B., Jr., W. R. Deebel, and R. D. Geisenderfer, eds. *Glossary of Oceanographic Terms*. 2nd ed. U.S. Naval Oceanographic Office Special Publication 35. Washington, D.C.: Government Printing Office, 1966.

King, C. A. M. *Beaches and Coasts*. New York: St. Martin's Press, 1972.

National Oceanographic and Atmospheric Administration, National Ocean Survey. *Our Restless Tides: A Brief Explanation of the Basic Astronomical Factors Which Produce Tides and Tidal Currents*. February 1998. Tides and Currents. http://www.co-ops.nos.noaa.gov/restles1.html (accessed November 7, 2009).

——. What are the "Perigean Spring Tides"? Do they cause coastal flooding? Tides and Currents. http://www.co-ops.nos.noaa.gov/faq2.html (accessed August 10, 2011).

Pilkey, O. H. *A Celebration of the World's Barrier Islands*. New York: Columbia University Press, 2003.

Pinet, P. R. *Invitation to Oceanography*. 3rd ed. Sudbury, Mass.: Jones and Bartlett, 2003. [Adapted from A. C. Redfield, *Introduction to Tides: The Tides and Waters of New England and New York* (Woods Hole, Mass.: Marine Science International, 1980)]

Walker, J. Lunar perigee and apogee calculator. 1997. http://www.fourmilab.ch/earth view/pacalc.html (accessed August 10, 2011).

West Nova Eco Site. Climate and Tides. http://www.collectionscanada.gc.ca/eppp-archive/100/200/301/ic/can_digital_collections/west_nova/climate.html (accessed September 5, 2011).

5. SEDIMENTS

Folk, R. L. *Petrology of Sedimentary Rocks*. Austin, Tex.: Hemphill, 1972.

Gary, M., R. McAfee Jr., and C. L. Wolf, eds. *Glossary of Geology*. Washington, D.C.: American Geological Institute, 1972.

Hjulstrom, F. Transportation of detritus by moving water. In *Recent Marine Sediments: A Symposium*, edited by P. D. Trask, 5–31. Tulsa, Okla.: American Association of Petroleum Geologists, 1939.

Pettijohn, F. J. *Sedimentary Rocks*. 2nd ed. New York: Harper & Row, 1957.

Reineck, H.-E., and I. B. Singh. *Depositional Sedimentary Environments, with Reference to Terrigenous Clastics*. New York: Springer-Verlag, 1980.

Shepard, F. P. Nomenclature based on sand-silt-clay ratios: An interim report. *Journal of Sedimentary Petrology* 24 (1954): 151–158.

Udden, J. A. Mechanical composition of clastic sediments. *Bulletin of the Geological Society of America* 25 (1914): 655–744.

——. *The Mechanical Composition of Wind Deposits*. Augustana Library Publication, no. 1. Rock Island, Ill.: Lutheran Augustana Book Concern, 1898.

Wentworth, C. K. A scale of grade and class terms for clastic sediments. *Journal of Geology* 30 (1922): 377–392.

6. BARRIER ISLANDS AND TIDAL INLETS

Davis, R. A., Jr., and D. M. FitzGerald. *Beaches and Coasts*. Malden, Mass.: Blackwell Science, 2004.

National Oceanographic and Atmospheric Administration. NOAA aircraft takes dramatic photos of North Carolina coast after Hurricane Isabel unleashed her fury. September 23, 2003. NOAA News Online. http://www.noaanews.noaa.gov/stories/s2091.htm; http://www.noaanews.noaa.gov/stories/images/hatteras-new-inlet-comparison.jpg (accessed August 8, 2011).

North Carolina Coastal Federation. Cape Hatteras Coast Keeper issues: Hurricane Isabel damage report. http://www.nccoast.org/ch-issues.htm (accessed February 25, 2005).

Pilkey, O. H. *A Celebration of the World's Barrier Islands*. New York: Columbia University Press, 2003.

Stewart, R. H. Coastal processes and tides. In *Introduction to Physical Oceanography*. 2005. Texas A&M University. http://oceanworld.tamu.edu/resources/ocng_textbook/chapter17/chapter17_01.htm (accessed August 8, 2011).

U.S. Army Corps of Engineers. Coastal morphodynamics. In *Engineering and Design: Coastal Geology*. Coastal Engineering Manual, EM 1110–2-1810, chap. 4. 1995. http://140.194.76.129/publications/eng-manuals/em1110-2-1810/c-4.pdf (accessed August 8, 2011).

——. Hurricane Isabel breaches Hatteras Island. Engineer Update, 27(11). 2003. http://www.hq.usace.army.mil/cepa/pubs/nov03/story8.htm (accessed February 25, 2005).

——, U.S. Army Engineer Research and Development Center. Coastal Inlets Research Program. Inlets Online. http://www.oceanscience.net/inletsonline/ (accessed August 8, 2011).

U.S. Geological Survey. Quinby Inlet NE DOQQ (Digital Orthophoto Quarter Quadrangle [infrared]). http://iris.lib.virginia.edu/mrsid/bin/get_image.pl?image=/lv5/siddoq/37075/37075d66.sid&size=thumbnail (accessed December 6, 2009).

A Virginia Gazetteer. University of Virginia Library. http://fisher.lib.virginia.edu/
collections/gis/vagaz/search_by_quad.php (accessed January 4, 2010).

Wright, L. D. *Morphodynamics of Inner Continental Shelves*. Boca Raton, Fla.: CRC Press, 1995.

Wright, L. D., and A. D. Short. Morphodynamic variability of surf zones and beaches: A synthesis. *Marine Geology* 56 (1984): 93–118.

7. SAND DUNES AND SALT MARSHES

Florida Oceanographic Society. Shoal grass (*Halodule wrightii*). http://www.florida
oceanographic.org/environ/seagrass7.htm (assessed December 7, 2009).

Shumway, S. W. *The Naturalist's Guide to the Atlantic Seashore: Beach Ecology from the Gulf of Maine to Cape Hatteras*. Guilford, Conn.: Globe Pequot Press / Falcon Guide, 2008.

U.S. Army Coastal Engineering Research Center. *Shore Protection Manual*. 3 vols. Fort Belvoir, Va.: U.S. Army Coastal Engineering Research Center, 1973.

8. SEA LEVEL AND SEA-LEVEL RISE

Belknap, D. F, B. G. Andersen, R. S. Anderson, H. W. Borns Jr., G. L. Jacobson, J. T. Kelley, R. C. Shipp, D. C. Smith, R. Stuckenrath Jr., W. B. Thompson, and D. A. Tyler. Late Quaternary sea-level changes in Maine. In *Sea-Level Fluctuations and Coastal Evolution*, edited by D. Nummendal, O. H. Pilkey, and J. D. Howard, 71–85. Special Publication, no. 41. Tulsa, Okla.: Society of Economic Paleontologists and Mineralogists, 1987.

Blanchon, P., and J. Shaw. Reef drowning during the last deglaciation: Evidence for catastrophic sea-level rise and ice-sheet collapse. *Geology* 23, no. 1 (1995): 4–8.

Bloom, A. L. Pleistocene shorelines: A new test of isostasy. *Geological Society of America Bulletin* 78, no. 12 (1967): 1477–1494.

Chappell, J., and N. J. Shackleton. Oxygen isotopes and sea level. *Nature* 324 (1986): 137–140.

Craft, C., J. Clough, J. Ehman, S. Joye, R. Park, S. Pennings, H. Guo, and M. Machmuller. Forecasting the effects of accelerated sea-level rise on tidal marsh ecosystem services. *Frontiers of Ecology and the Environment* 7, no. 2 (2009): 73–78.

Douglas, B. C. Sea level change in the era of the recording tide gauge. In *Sea Level Rise: History and Consequences*, edited by B. C. Douglas, M. S. Kearney, and S. P. Leatherman, 37–64. San Diego, Calif.: Academic Press, 2001.

Gary, M., R. McAfee Jr., and C. L. Wolf, eds. *Glossary of Geology*, Washington, D.C.: American Geological Institute, 1972.

Gornitz, V. *Rising Seas: Past, Present, Future*. New York: Columbia University Press, 2013.

Jevrejeva, S., J. C. Moore, and A. Grinsted. How will sea level respond to changes in natural and anthropogenic forcings by 2100? *Geophysical Research Letters* 37 (2010): L07703.

Kaye, C. A., and E. S. Barghoorn. Late Quaternary sea-level change and crustal rise at Boston, Massachusetts, with notes on the autocompaction of peat. *Geological Society of America Bulletin* 75, no. 2 (1964): 63–80.

Kelley, J. T., S. M. Dickson, and D. F. Belknap. Maine's history of sea-level changes. 2005. Main Geological Survey. http://www.maine.gov/doc/nrimc/mgs/explore/ marine/facts/sealevel.htm (accessed January 8, 2010).

Kelley, J. T., S. M. Dickson, D. F. Belknap, and R. Stuckenrath Jr.. Sea-level change and late Quaternary sediment accumulation on the southern Maine inner continental shelf. In Quaternary Coasts of the United States: Marine and Lacustrine Systems, edited by C. H. Fletcher III and J. F. Wehmiller, 23–34. Special Publication, no. 48. Tulsa, Okla.: Society of Economic Paleontologists and Mineralogists, 1992.

Koeteff, C., J. R. Stone, F. D. Larsen, and J. H. Hartshorn. Glacial Lake Hitchcock and Post-glacial Uplift. Open File Report OF 87–0329. Washington, D.C.: U.S. Geological Survey, 1987.

Lian, O. B., and R. G. Roberts. Dating the Quaternary: Progress in luminescence dating of sediments. Quaternary Science Reviews 25 (2006): 2449–2468.

Liu, J. P. Post-glacial sedimentation in a river-dominated epicontinental shelf: The Yellow Sea example. Ph.D. diss., Virginia Institute of Marine Science, College of William & Mary, 2001.

Liu, J. P., J. D. Milliman, S. Gao, and P. Cheng. Holocene development of the Yellow River's subaqueous delta, North Yellow Sea. Marine Geology 209 (2004): 45–67.

Maine Geological Survey. Bucksport-Searsport region feels small earthquake swarm. Press release, May 3, 2011. Maine Geological Survey. http://www.maine.gov/doc/ nrimc/mgs/explore/hazards/quake/bucksport-microquakes-press.htm (accessed May 20, 2011).

Milliman, J. D., and K. O. Emery. Sea levels during the past 35,000 years. Science 162 (1968): 1121–1123.

National Oceanic and Atmospheric Administration. Sea Levels Online: Sea Level Trends. http://co-ops.nos.noaa.gov/sltrends/sltrends.shtml (accessed June 12, 2011).

——, National Ocean Service, Center for Operational Oceanographic Products and Services. Tidal Datums and Their Applications. NOAA Special Publication NOS CO-OPS 1. 2000. http://tidesandcurrents.noaa.gov/publications/tidal_datums_and_their_ applications.pdf (accessed January 8, 2010).

Proudman Oceanographic Laboratory. http://www.pol.ac.uk (accessed August 15, 2011).

Shaw, J. The meltwater hypothesis for subglacial bedforms. Quaternary International 90 (2002): 5–22.

Solomon, S., D. Qin, M. Manning, Z. Chen, M. Marquis, K. B. Averyt, M. Tignor, and H. L. Miller, eds. Climate Change 2007: The Physical Science Bases: Contribution of Working Group 1 to the Fourth Assessment Report of the Intergovernmental Panel on Climate Change. Cambridge: Cambridge University Press, 2007. Also available at http://www.ipcc .ch/ipccreports/ar4-wg1.htm (accessed September 28, 2010).

University of Oxford, School of Archaeology. Luminescence dating. http://www.arch .ox.ac.uk/L.html (accessed August 15, 2011).

University of Washington, Luminescence Dating Laboratory. Luminescence dating in brief. http://depts.washington.edu/lumines/ (accessed January 8, 2010).

U.S. Geological Survey. Sea level and climate change. USGS Fact Sheet 002-00. 2000. http://pubs.usgs.gov/fs/fs2-00/ (accessed January 8, 2010).

Vermeer, M., and S. Rahmstorf. Global sea level linked to global temperature. *Proceedings of the National Academy of Sciences of the United States of America* 106, no. 51 (2009): 21527–21532.

Williams, R. S., and D. K. Hall. Glaciers. In *Atlas of Earth Observations Related to Global Change*, edited by R. J. Gurney, J. L. Foster, and C. L. Parkinson, 401–422. Cambridge: Cambridge University Press, 1993.

9. STORMS AND STORM SURGE

Allen, E. S. *A Wind to Shake the World: The Story of the 1938 Hurricane.* Boston: Little, Brown, 1976.

Hobbs, C. H., III. Shoreline orientation and storm surge. *Maritime Sediments* 6, no. 3 (1970): 113–115.

Junger, S. *The Perfect Storm: A True Story of Men Against the Sea.* New York: Norton, 1997.

Knabb, R. D., J. R. Rhome, and D. P Brown. Tropical cyclone report, Hurricane Katrina, 23–30 August 2005. December 20, 2005. National Weather Service, National Hurricane Center. http://www.nhc.noaa.gov/pdf/TCR-AL122005_Katrina.pdf (accessed August 15, 2011).

Larson, E. . *Isaac's Storm: A Man, a Time, and the Deadliest Hurricane in History.* New York: Crown, 1999.

Morton, R., K. Guy, and H. Hill. Morphological impacts of the March 1962 storm on barrier islands of the Middle Atlantic states. Last modified March 10, 2010. U.S. Geological Survey, St. Petersburg Coastal and Marine Sciences Center. http://coastal.er.usgs.gov/hurricanes/historical-storms/march1962 (accessed August 15, 2011).

National Oceanic and Atmospheric Administration. Historic tide data. Tides and Currents. http://tidesandcurrents.noaa.gov/station_retrieve.shtml?type=Historic+Tide+Data (accessed January 4, 2010).

——, National Climatic Data Center. Hurricane Katrina. Last modified December 29, 2005. http://www.ncdc.noaa.gov/oa/climate/research/2005/katrina.html (accessed April 2006).

Sallenger, A. *Island in a Storm: A Rising Sea, a Vanishing Coast, and a Nineteenth-Century Disaster that Warns of a Warmer World.* New York: Public Affairs, 2009.

Scotti, R. A. *Sudden Sea: The Great Hurricane of 1938.* Boston: Little, Brown, 2003.

Storm surge, *Wikipedia.* Last modified July 30, 2011. http://en.wikipedia.org/wiki/Storm_surge (accessed August 15, 2011).

10. EROSION AND SHORE PROTECTION

Clark, I., C. E. Larsen, and M. Herzog, 2004. Evolution of equilibrium slopes at Calvert Cliffs, Maryland: A method of estimating the timescale of slope stabilization. *Shore & Beach* 72, no. 4 (2004): 17–23.

Davis, R. A., Jr., and D. M. FitzGerald. *Beaches and Coasts*. Malden, Mass.: Blackwell Science, 2004.

Dean, R. G. *Beach Nourishment: Theory and Practice*. Advanced Series on Ocean Engineering, vol. 18. River Edge, N.J.: World Scientific, 2002.

Dean, R. G., R. A. Davis, and K. M. Erickson. Beach nourishment with emphasis on geological characteristics affecting project performance. In *Beach Nourishment: A Guide for Local Government Officials*. National Oceanic and Atmospheric Administration, Coastal Services Center. http://www3.csc.noaa.gov/beachnourishment/html/geo/scitech.htm (accessed March 1, 2005).

Douglass, S. L. *Saving America's Beaches: The Causes of and Solutions to Beach Erosion*. Advanced Series on Ocean Engineering, vol. 19. River Edge, N.J.: World Scientific, 2002.

Finkl, C. W., and C. H. Hobbs III. Mining sand on the continental shelf of the Atlantic and Gulf coasts of the U.S. *Marine Georesources and Geotechnology* 27 (2009): 230–253.

U.S. Army Corps of Engineers, U.S. Army Engineer Research and Development Center, Coastal Inlets Research Program. Inlets Online. http://www.oceanscience.net/inletsonline (accessed August 8, 2011).

INDEX

Numbers in italics refer to pages on which figures appear.